Materials Horizons: From Nature to Nanomaterials

Series Editor

Vijay Kumar Thakur, School of Aerospace, Transport and Manufacturing, Cranfield University, Cranfield, UK

Materials are an indispensable part of human civilization since the inception of life on earth. With the passage of time, innumerable new materials have been explored as well as developed and the search for new innovative materials continues briskly. Keeping in mind the immense perspectives of various classes of materials, this series aims at providing a comprehensive collection of works across the breadth of materials research at cutting-edge interface of materials science with physics, chemistry, biology and engineering.

This series covers a galaxy of materials ranging from natural materials to nanomaterials. Some of the topics include but not limited to: biological materials, biomimetic materials, ceramics, composites, coatings, functional materials, glasses, inorganic materials, inorganic-organic hybrids, metals, membranes, magnetic materials, manufacturing of materials, nanomaterials, organic materials and pigments to name a few. The series provides most timely and comprehensive information on advanced synthesis, processing, characterization, manufacturing and applications in a broad range of interdisciplinary fields in science, engineering and technology.

This series accepts both authored and edited works, including textbooks, monographs, reference works, and professional books. The books in this series will provide a deep insight into the state-of-art of Materials Horizons and serve students, academic, government and industrial scientists involved in all aspects of materials research.

Review Process

The proposal for each volume is reviewed by the following:

1. Responsible (in-house) editor
2. One external subject expert
3. One of the editorial board members.

The chapters in each volume are individually reviewed single blind by expert reviewers and the volume editor.

Arnab Chanda · Gurpreet Singh

Mechanical Properties of Human Tissues

Springer

Arnab Chanda
Centre for Biomedical Engineering
Indian Institute of Technology Delhi
New Delhi, India

Gurpreet Singh
Centre for Biomedical Engineering
Indian Institute of Technology Delhi
New Delhi, India

ISSN 2524-5384 ISSN 2524-5392 (electronic)
Materials Horizons: From Nature to Nanomaterials
ISBN 978-981-99-2224-6 ISBN 978-981-99-2225-3 (eBook)
https://doi.org/10.1007/978-981-99-2225-3

This Springer imprint is published by the registered company Springer Nature Singapore Pte Ltd.
The registered company address is: 152 Beach Road, #21-01/04 Gateway East, Singapore 189721, Singapore

Preface

The book titled *Mechanical Properties of Human Tissues* discusses the human soft tissues from head to toe, under one umbrella, and presents their mechanical properties which can be beneficial for applications in engineering, biosciences, biomedical, and tissue engineering. To date, numerous experimental researches have been conducted to estimate human tissue properties, but these exercises have taken place in silos. Majorly, the biomechanical properties of peripheral tissues like skin have been studied widely. However, only a few studies have investigated the properties of internal tissues such as the brain and the heart. This work will be the first-of-its-kind to bring all such research methods and outcomes of human tissue experiments under a single umbrella. Different mechanical characterization techniques employed in human tissue property estimation are presented in detail. Especially, hyperelastic constitutive models (e.g., Mooney–Rivlin, Ogden) for both isotropic and anisotropic tissues are summarized. Human tissues, including the skin, muscles, connective tissues, and tissues in all functional organs, have been listed, and their mechanical properties are presented in detail. The mechanical properties of human tissues are also crucial in various key areas, including trauma research, ballistic testing, surgical planning, and the mitigation of human injuries (e.g., Armor development).

This book is divided into eight chapters that cover different human tissues from head to toe. Chapter 1 gives an overview of soft human tissues and presents them under four categories, i.e., skin, muscles, connective tissues, and functional organs. Further, the mechanical characterization techniques, both linear and nonlinear methods, are discussed in detail. Skin is the largest organ on the human body, and Chap. 2 is fully dedicated to the skin tissue. It covers the structure, mechanical properties, and frictional properties of the skin in detail. The muscles and connective tissues are responsible for the movement of the human body to perform our daily routines. Chapter 3 deals with the mechanical properties of these tissues, i.e., muscles (skeletal, smooth, and cardiac) and connective tissues (ligaments and tendons). Chapters 4–6 cover the tissues in the functional organs. Chapter 4 covers the tissues in functional organs with low stiffness, i.e., brain, tongue, tonsils, oesophagus, lungs, breast, stomach, and spleen. Chapter 5 covers the tissues in functional organs with medium stiffness, i.e., liver, gallbladder, kidney, and uterus. The remaining tissues are categorized under tissues in functional organs with high stiffness and includes the nasal cavity, oral cavity, heart, pancreas, small intestine, colon, vagina, and urinary bladder. Chapter 7 presents the hyperelastic models for anisotropic tissue characterization including different numerical models for modeling the effect of fiber and matrix contributions, modeling the effect of fiber orientation, and the effect of multiple fiber layers. Finally, Chap. 8 discusses the challenges and applications of the presented work. In addition, the key future opportunities were also highlighted including the measurement of mechanical properties of internal organs in normal, diseased, and damaged conditions, characterizing tissue anisotropy, handling and management of tissues, and ethical issues with in-vivo testing.

The knowledge of the mechanical properties of the human tissues would be indispensable for healthcare researchers, biomedical and biotech engineers, research in tissue engineering, medical simulation and devices, computational and finite element modeling (FEM), defence research (ballistics, armor development), biomimetic applications, etc. This book can be used as a comprehensive source for scientists, academicians, researchers, and engineers in various areas, and we are highly confident that this contribution will benefit all the readers in different ways.

New Delhi, India
March 2023

Arnab Chanda
arnab.chanda@cbme.iitd.ac.in

Gurpreet Singh
gurpreet.singh@cbme.iitd.ac.in

Contents

About the Authors

Dr. Arnab Chanda is an Assistant Professor in the Centre for Biomedical Engineering, Indian Institute of Technology (IIT) Delhi, India and a joint faculty at the Department of Biomedical Engineering, All India Institute of Medical Sciences (AIIMS) Delhi, India. He is also the founder of a startup company BIOFIT Technologies LLC, USA.

Dr. Chanda is an expert in the fabrication and mechanical characterization of tissue mimics, and has previously developed artificial surrogates for human skin, muscles, brain, artery, and plantar fascia, and tested them at both lab and clinical settings.

These experimental models have been used extensively for surgical training and to study a wide range of injury scenarios. To date, he has received young researcher awards from ASME and MHRD, holds 7 US patents, 2 Indian patents, has authored over 50 articles in reputed international journals. Currently, Dr. Chanda heads the "Disease and Injury Mechanics Lab (DIML)", where his team is working on developing cutting-edge healthcare technologies for disease mitigation (i.e., diabetic ulceration, cerebral aneurysm, and severe skin burns) in India. They also aim to fabricate low-cost artificial organs for surgical training.

Gurpreet Singh is a Ph.D. Scholar in the Centre for Biomedical Engineering, Indian Institute of Technology (IIT) Delhi, India. He is a recipient of the most prestigious Ph.D. fellowship in India, the Prime Minister's Research Fellowship (PMRF), in the May 2021 cycle.

His research interests are soft tissue mechanics, artificial tissues, biomimetics, and computational biomechanics. He is currently developing artificial human tissues for injury and disease modeling.

Previously, much of his work has been on improving the surface characteristics and bioactivity of metallic biomaterials, with research interests including surface engineering, materials science, biomaterials, and non-conventional machining processes. He worked on the surface modification of metallic biomaterials using electro-discharge machining, where he studied the bioactivity of modified surfaces in terms of wear resistance, corrosion resistance, and other biological responses. He has contributed 34 research papers and book chapters to leading international journals/conferences. He is also serving as a reviewer for prominent journals of national and international repute.

Chapter 1
Introduction to Human Tissues

1.1 Introduction

Soft tissues are made up of multiple fibrous layers and embedded in a gelatinous matrix. These are flexible tissues and are not hardened by the calcification process such as hard tissues (e.g., bones). Soft tissues exhibit anisotropic behavior because of their fiber orientations and shows non-homogeneous material properties at microscopic level. The stress–strain responses of the soft tissues are nonlinear in behavior and significantly depends on the applied strain rate. Unlike hard tissues, the soft tissues possess large deformation during the mechanical testing. In addition, soft tissues exhibit viscoelastic behavior due to the shear interaction between collagen and proteoglycan matrix. Soft tissues can be broadly divided in four categories as follows:

- Skin
- Muscles
- Connective tissues
- Functional organs

The skin, which acts as a first contact or barrier for the external stimuli, is the largest soft tissue on the human body [1, 2]. The muscles can be categorized as skeletal muscles, smooth muscles, and cardiac muscles. In general, the muscles are responsible for providing the support and stability to the skeleton and give it power to move. Tendons and ligaments are two types of connective tissues which provide mobility to the human body by transferring the loads between bone to bone and bone to muscles. The functional organs cover a wide range of soft tissues across the length and width of the human body (Fig. 1.1). The major soft tissues include brain, oral cavity, nasal cavity, tongue, tonsils, salivary glands, esophagus, lungs, heart, breast, liver, gallbladder, stomach, small intestine, colon, spleen, pancreas, kidney, urinary bladder, uterus, and vagina.

A. Chanda and G. Singh, *Mechanical Properties of Human Tissues*,
Materials Horizons: From Nature to Nanomaterials,
https://doi.org/10.1007/978-981-99-2225-3_1

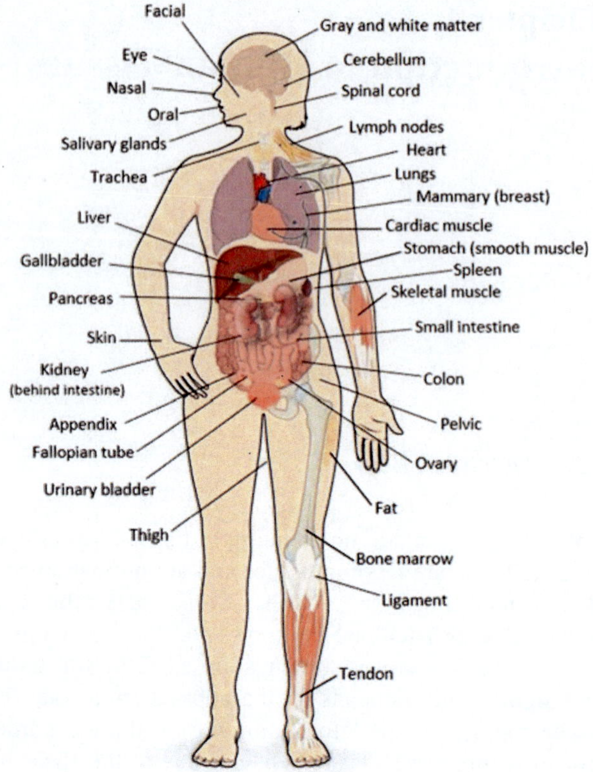

Fig. 1.1 Soft tissues located across the human body [3]

The brain tissue, which is central nervous system of the human body, is a wrinkled organ located within the cranial cavity. It transmits the information to the body and is composed of the intracranial cerebrum and the cerebellum, and the spinal cord [4, 5]. The nasal cavity is the first airway pass and is the primary external opening of the respiratory system. The function of the nasal cavity is to warm, moisturize, and filter the air before it reaches the lungs. In addition to the nasal cavity, the oral cavity is a secondary path to the respiratory system. It is shorter air pathway than the nasal cavity and advantageous to facilitate the fast reaching of air to the lungs [3]. The tongue is a functional organ of the upper airway, aiding in swallowing, speaking, licking, and breathing [6]. The taste buds on the surface of tongue detect the flavor of food, and the information is then passed to the brain [7, 8]. The tonsils are a part of our immune system, which restricts the entry of bacteria and viruses in our body to prevent infection in the throat and lungs. These are the freshly lumps presented on each side of the mouth.

The lung tissue is a pair of big, spongy organs placed lateral to the heart and superior to the diaphragm in the thorax. The lung's primary work is to provide oxygen to the blood and to cleanse it of carbon dioxide in exchange (via tiny air sacs called alveoli). The heart is a pumping organ located in the thoracic area of the human body. It is positioned medial to the lungs and to the body's midline. The liver, which is the second largest organ after the skin, is located on the right side of the abdomen. The primary function of the liver tissue is to regulate the metabolism of proteins, carbohydrates, and lipids and the storage of minerals and vitamins [9, 10]. The gallbladder tissue is a thin pear-shaped organ connected directly under the right lobe of the liver. Its primary function is to store and release bile into the duodenum for the proper digestion of the meal using different lipids [11]. The esophagus is a muscular tube which delivers food and water to the stomach through the neck region (pharynx). At the inferior end of esophagus, the cardiac sphincter seals the esophagus and captures food in the stomach [12, 13].

The stomach is a major organ of the gastrointestinal system and acts as a store for nutrients. It is a muscular sac located in the upper abdomen between the esophagus and the duodenum [14]. Small intestine is part of the lower gastrointestinal tract, which includes the duodenum, jejunum, and ileum [15]. It is positioned under the stomach and occupies the maximum abdominal cavity. The small intestine is coiled like a hose and promotes digestion and absorption of nutrients [16]. The colon, also known as the large intestine, is slightly inferior to the stomach and coils along the top and lateral edge of the small intestine. The spleen is a flattened, oval-shaped soft tissue located in the upper left abdomen, lateral to the stomach. The spleen has a large blood reserve that functions to filter large amounts of blood [17]. The pancreas tissue is positioned just inferior and posterior to the stomach in the abdominal cavity. It shapes like a lumpy snake with its "head" linked to the duodenum and "tail" pointing to the left wall of the abdominal cavity. The pancreas secretes digestive enzymes into the small intestine for the chemical breakdown of carbohydrates. The kidney is bean-shaped tissue positioned along the posterior wall of the abdominal cavity. The kidneys are in contact with the back muscles and are located posterior to the peritoneum. A thin layer of adipose protects the kidneys, keeps them in place, and prevents physical damage. The urinary bladder, located near the pelvis base and in the midline of the body, is a hollow organ that collects and excretes urine [18]. The vaginal tissue is an elastic and muscular tube and acts as a delivery channel for childbirth and pathway for the menstrual blood [19, 20]. The vaginal tissue is about 7.5 cm along the anterior wall and 9 cm along the posterior wall [21]. The vaginal tissue is connected to the uterus tissue and fallopian tubes through the cervix [21, 22].

Soft tissues such as skin, vascular tissues, neural tissues, skeletal muscle, cartilage, and ligaments lose their normal functionality due to pathological diseases, cancer, or trauma. Such soft tissues need restoration or improvement for the adequate functionality of the damaged tissue [23]. However, the functioning should also consider the chemical and biological properties of the tissue along with its mechanical properties. In addition, the properties of the new material should be fully biocompatible. Polymers have been extensively studied as a tissue replacement biomaterial for fabricating soft tissues, such as skin, heart valves, nerves, arteries, liver, pancreas, bladder, and other soft tissues [23–28]. Polymers such as collagen, alginate, elastin, chitosan, polylactide (PLA), polyglycerol-sebacate (PGS), poly (lactic-co-glycolic acid) (PLGA), and polyurethane (PU) are some natural polymers which are extensively used by the researchers to mimic the mechanical properties of the natural soft tissues [29–33]. Other than the polymers, hydrogels are renowned for the tissue engineering considering their prominence of mimicking the realistic mechanical properties of the human soft tissues [34–36].

The inherent properties or behavior of the soft tissues significantly varies depending on the environment they are existing (such as body fluids, enzymes, blood flow), which also affects their mechanical properties. The mechanical properties of such soft tissues can be measured or estimated using different in vivo and ex vivo measuring techniques. Magnetic resonance elastography and ultrasonic elastography are two commonly used in vivo techniques for evaluating the mechanical properties of soft tissues [37]. The working principle of these techniques is to measure the induced vibrations and displacement due to indentation and compression on the targeted soft tissue. To assess the mechanical properties of peripheral tissues such as forearms, thighs, forehead skin, plantar foot tissues, and spinal tissues, the indentation is the most prominently used technique [38, 39]. In addition, in vivo indentation has been extensively utilized to evaluate the mechanical properties of the brain [40], heart [41], liver [42], breast [43], skeletal muscle [44, 45], and lung tissue.

Ethical considerations, accessibility and availability of tissue samples, repeatability of results across samples, and varying boundary conditions make it difficult to experimentally test the cadaveric tissues for ex vivo testing [46]. For the ex vivo testing, the tissue samples excised during the autopsy and tested in laboratory settings similar to the engineering materials using measuring methods such as uniaxial stretch, biaxial stretch, shear, and indentation. In some cases, the tissue samples may be from surgical leftovers or an animal model (porcine, bovine, etc.).

1.2 Mechanical Characterization Techniques

1.2.1 Linear

Tensile Testing

Across all the linear mechanical characterization techniques for soft tissues, uniaxial and biaxial testing are standard techniques for determining their mechanical properties. The most well-known method for analyzing the mechanical characteristics of different materials is tensile testing [47] and has also been employed to estimate the mechanical properties of soft tissues [48, 49]. The procedure involves fixing the sample's ends with a fixture and pushing it apart until it fails or reaches a specific strain (Fig. 1.2). The sample's stress resulting from the external load is determined using the sample's cross-sectional area and the tensile force applied to it. Stress–strain curves are created by continually recording the sample's elongation and the computed stress throughout the test. Plotting the stress–strain curves allows for the extraction of many mechanical parameters, including the Young's modulus, yield strength, and ultimate tensile strength. In order to strengthen the hold of their tensile

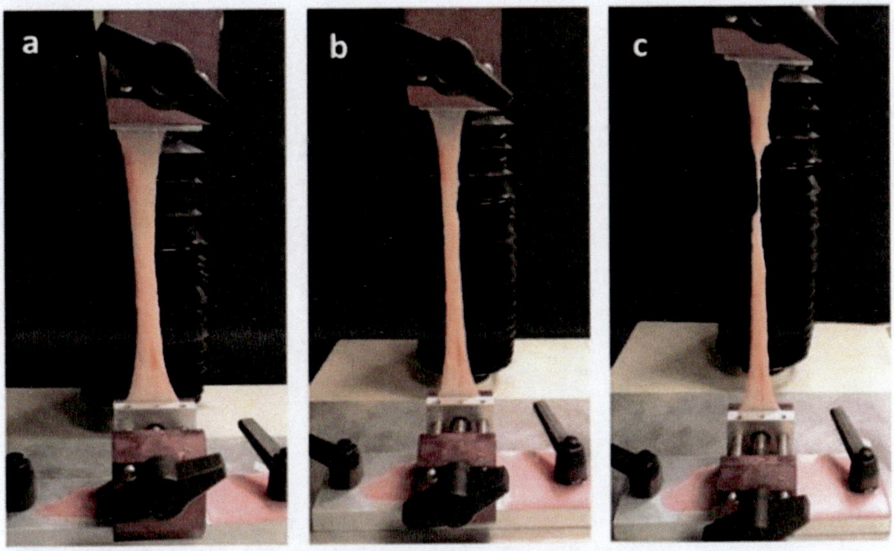

Fig. 1.2 Uniaxial testing of the porcine bladder tissue in the longitudinal direction (loaded up to rupture from **a** to **c**) [51]

testing system, Scholze et al. [47] used 3D-printed clamps and used their design on human skin, ligament, and tendon tissues. In the recent years, a new technique has been developed that uses a camera to automatically track the movements of painted markers on the sample in a contactless manner to quantify strain. The technique's feasibility was examined using uniaxial tensile testing on human and pig tendon tissues. The outcomes showed an improvement in precision and the capacity to classify various areas of a diverse sample.

The stress–strain curve can be used to directly determine mechanical parameters without the use of extra models, and the tensile testing method does not require a particularly complex setup. Its remarkable reproducibility and adaptability, such as the capacity to include various axes of stress during trials, are also a result of this simple procedure. The tensile testing method also has some issues; for example, a significant drawback of the method is material slippage. The sample must be exactly aligned with the loading axis since gravity factors may cause the sample to stray from this axis. Also, sample preparation can be challenging, particularly when it comes to tissues with irregular geometries that are challenging to cut into the right shape for tensile testing. During uniaxial tensile testing, the tissue can be stretched while the unconfined edges can contract and the fiber can be crushed in the lateral direction [50]. These drawbacks can be overcome by biaxial tensile testing, which eliminates fiber compression and more accurately simulates the in situ loading of fibrous tissues [50]. In general, the uniaxial loading is preferred to estimate the isotropic behavior of the soft tissues; however, the biaxial testing simultaneously stretches the samples

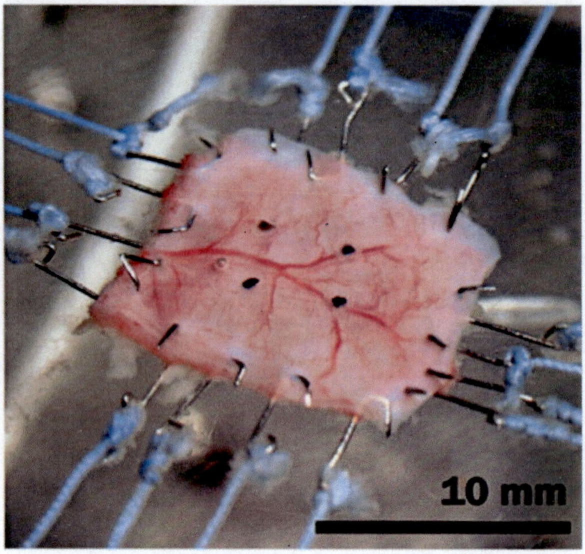

Fig. 1.3 Biaxial testing of the porcine small intestine in the circumferential and longitudinal directions [15]

in both the directions (e.g., skin) and provides the anisotropic properties of a tissue sample (Fig. 1.3).

Compression Testing

Another common characterization method is compression testing, which is similar to tensile testing except that the sample is compressed rather than stretched under tension (Fig. 1.4). It operates on the same fundamental tensile testing premise, measuring the force applied to the sample to determine its stress. Stress–strain curves are plotted by continuously recording the sample's stress and strain. However, the sample is positioned between two flat platens for compressive loading rather of being clamped. Contained compression and unconfined compression are the two fundamental methods by which compression testing can be carried out. When a sample is squeezed with a porous platen inside of an impermeable well, it allows fluid to be shifted vertically from within the sample [52]. Unconfined compression causes fluid to flow radially by compressing the sample between an impermeable base and a non-porous platen [52].

Numerous soft tissue studies have been characterized using compression testing. Using compression testing, Wu et al. [54] developed a novel method for assessing the nonlinear-elastic behavior of skin and subcutaneous tissue without separating them. They experimented using composite samples acquired from pig foot to validate their method. In order to determine the friction coefficients during compression testing on swine brain tissue, Rashid et al. [55] used a hybrid experimental–computational method. The experimental stress data was utilized as a reference for the numerical simulations to estimate the friction coefficients. The brain samples were compressed at various strain rates and held between metal platens in either a bonded (no slip) or lubricated (pure slip) condition.

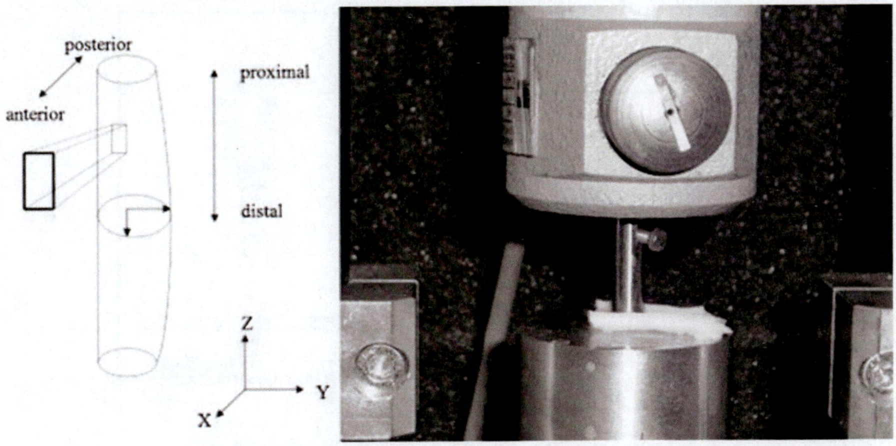

Fig. 1.4 Set up to compress healthy, calcified, diseased porcine and human aorta tissue [53]

The most prominent benefit of compression testing is that it evaluates soft tissues under physiologically realistic stresses in a moist environment [52]. The accuracy of the mechanical properties evaluated is improved by testing these tissues under compression because many load-bearing tissues throughout the body encounter compressive stresses. Similar to tensile testing, compression testing offers the benefit of a simple method to extract parameters directly from the stress–strain curves. The approach has some drawbacks, including challenges with sample preparation. Samples must be cut to exacting measurements, and their planes must be free of any surface imperfections. The accuracy of the stress–strain recordings can be significantly impacted by friction between both the surface of the platen and the sample, which results in an overestimation of stress.

Indentation Testing

By using a small indenter to examine the sample's surface, indentation can be used to determine the mechanical properties of a material (Fig. 1.5). Force–displacement curves are plotted by recording the external force exerted on the sample during indentation and the displacement of the indenter during the experiment. In order to ascertain the mechanical characteristics of the sample, these data curves are fitted to a suitable indentation model. They strongly rely on the geometry of the indenter tip, and the ratio of the indenter's dimensions to the sample's dimensions has a significant impact on the models as well. Griffin et al. [52] suggested employing indentation testing as a standardized non-destructive method for identifying skin and cartilage. They tested their method by using a semi-circular indenter to measure the elastic and viscoelastic characteristics of samples of human cartilage.

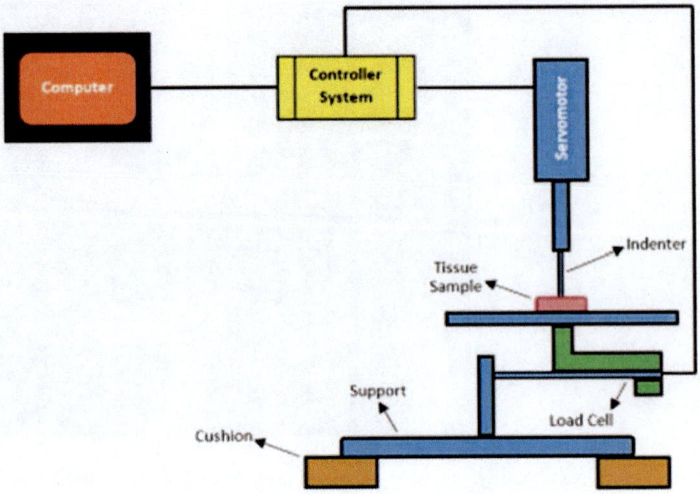

Fig. 1.5 Schematic of the tissue indentation measurement system [56]

1.2.2 Nonlinear

Under large strains, the theory of nonlinear elasticity can be used to characterize the mechanical properties of soft tissues. The anisotropic behavior of the soft tissues is generally estimated on the basis of material form-based strain-energy function (ψ). This strain-energy function of the hyperelastic model depends on Cauchy–Green tensor invariants (I_1, I_2, and I_3) or main stretches (λ_1, λ_2, and λ_3) (Eqs. 1.1 and 1.2).

$$\psi_{\text{Isotropic}} = \psi(I_1, I_2, I_3) \tag{1.1}$$

$$I_1 = \sum_{i=1}^{3} \lambda_i^2,\ I_2 = \sum_{i,j=1}^{3} \lambda_i^2\lambda_j^2,\ I_3 = \prod_{i=1}^{3} \lambda_i^2 \tag{1.2}$$

The Mooney–Rivlin model was one of the early isotropic hyperelastic models. The mechanical test data of nonlinear materials, such as rubber and soft tissues, under uniaxial and biaxial stresses, was accurately fitted using this method. Equation 1.3 may curve-fit the stress vs stretch data and determine the coefficients c_1 and c_2 with a stretch of "λ" applied uniaxially. The Mooney–Rivlin model, which was used to properly define the nonlinear behavior of soft materials like the breast and brain tissues, has been reduced to a polynomial form in the Yeoh model (Eq. 1.4).

$$\sigma_{\text{Mooney}} = 2\left(\lambda^2 - \frac{1}{\lambda}\right)\left(c_1 + c_2\frac{1}{\lambda}\right) \tag{1.3}$$

$$\sigma_{\text{Yeoh}} = 2\left(\lambda^2 - \frac{1}{\lambda}\right)\left(c_1 + 2c_2(I_1 - 3) + 3c_3(I_1 - 3)^2\right) \tag{1.4}$$

Neo-Hookean model was basically developed to estimate the mechanical behavior of the rubber subjected to uniaxial strain < 0.2 (Eq. 1.5). The Ogden model (Eq. 1.6) presented in Eq. 1.6 serves to be a reliable mechanical characterization model. For isotropic soft materials, the Ogden model is preferred where the mechanical properties are not affected by changing the strain rates (Eq. 1.6).

$$\sigma_{\text{Hookean}} = 2\left(\lambda^2 - \frac{1}{\lambda}\right)c_1 \tag{1.5}$$

$$\sigma_{\text{Ogden}} = c_1\left(\lambda^{c_2} - 2^{-1+c_2}\lambda^{-c_2/2}\right) + c_3\left(\lambda^{c_4} - 2^{-1+c_4}\lambda^{-c_4/2}\right) + c_5\left(\lambda^{c_6} - 2^{-1+c_6}\lambda^{-c_6/2}\right) \tag{1.6}$$

To evaluate the material property of soft tissue fibers, Humphrey and Yin derived composite-based hyperelastic model tissue fibers (Eq. 1.7).

$$\sigma_{\text{Humphrey}} = 2(\lambda^2 - \frac{1}{\lambda})c_1c_2e^{c_2(I_1-3)} \tag{1.7}$$

The knowledge of the mechanical properties of soft tissues is important for a wide range of applications such as surgical training, diagnostic purposes, ballistic testing, finite element modeling and fabricating the tissue simulants with realistic mechanical properties. These applications need biomechanical models having accurate geometry and realistic mechanical properties [46, 57, 58], and specifically, the mechanical properties of the soft tissues are used widely in numerical simulations to investigate the influence of various loading conditions on the stresses and rupture risk of soft tissues [59].

References

1. McGrath JA, Uitto J (2010) Anatomy and organization of human skin. Rook's Textb Dermatol 1: 1–53. (Wiley-Blackwell, Oxford). https://doi.org/10.1002/9781444317633.ch3
2. Chanda A, Unnikrishnan V, Lackey K (2017) Biofidelic conductive synthetic skin composites. In: 32nd Tech. Conference on American society for composites, vol 1, pp 400–8 (DEStech Publications Inc.). https://doi.org/10.12783/asc2017/15197
3. Singh G, Chanda A (2021) Mechanical properties of whole-body soft human tissues: a review. Biomed Mater 16:062004. https://doi.org/10.1088/1748-605X/AC2B7A
4. Budday S, Sommer G, Birkl C, Langkammer C, Haybaeck J, Kohnert J et al (2017) Mechanical characterization of human brain tissue. Acta Biomater 48:319–340. https://doi.org/10.1016/j.actbio.2016.10.036
5. Budday S, Sarem M, Starck L, Sommer G, Pfefferle J, Phunchago N et al (2020) Towards microstructure-informed material models for human brain tissue. Acta Biomater 104:53–65. https://doi.org/10.1016/j.actbio.2019.12.030

6. Hermant N, Perrier P, Payan Y (2017) Human tongue biomechanical modeling. Biomechanics of living organs. Hyperelastic constitutive laws for finite element modeling, pp 395–411. (Elsevier Inc.). https://doi.org/10.1016/B978-0-12-804009-6.00019-5
7. Kajee Y, Pelteret J-PV, Reddy BD (2013) The biomechanics of the human tongue. Int J Numer Method Biomed Eng 29:492–514. https://doi.org/10.1002/cnm.2531
8. Payan Y, Bettega G, Raphaël B (1998) A biomechanical model of the human tongue and its clinical implications. Lect Notes Comput Sci (including Subser Lect Notes Artif Intell Lect Notes Bioinform) 1496. Springer, Berlin, pp 688–695. https://doi.org/10.1007/bfb0056255
9. Kazemnejad S (2009) Hepatic tissue engineering using scaffolds: state of the art. Avicenna J Med Biotechnol 1:135–145
10. Lorenzini S, Andreone P (2007) Stem cell therapy for human liver cirrhosis: a cautious analysis of the results. Stem Cells 25:2383–2384. https://doi.org/10.1634/stemcells.2007-0056
11. Li WG, Hill NA, Ogden RW, Smythe A, Majeed AW, Bird N et al (2013) Anisotropic behaviour of human gallbladder walls. J Mech Behav Biomed Mater 20:363–375. https://doi.org/10.1016/j.jmbbm.2013.02.015
12. Mir M, Ali MN, Ansari U, Sami J (2016) Structure and motility of the esophagus from a mechanical perspective. Esophagus 13:8–16. https://doi.org/10.1007/s10388-015-0497-1
13. Hajhosseini P, Takalloozadeh M (2019) An isotropic hyperelastic model of esophagus tissue layers along with three-dimensional simulation of esophageal peristaltic behavior. J Bioeng Res 1:12–27. https://doi.org/10.22034/jbr.2019.189018.1009
14. Brandstaeter S, Fuchs SL, Aydin RC, Cyron CJ (2019) Mechanics of the stomach: a review of an emerging field of biomechanics. GAMM-Mitt 42. https://doi.org/10.1002/gamm.201900001
15. Bellini C, Glass P, Sitti M, Di Martino ES (2011) Biaxial mechanical modeling of the small intestine. J Mech Behav Biomed Mater 4:1727–1740. https://doi.org/10.1016/j.jmbbm.2011.05.030
16. Miyasaka EA, Okawada M, Utter B, Mustafa-Maria H, Luntz J, Brei D et al (2010) Application of distractive forces to the small intestine: defining safe limits. J Surg Res 163:169–175. https://doi.org/10.1016/j.jss.2010.03.060
17. Kemper AR, Santago AC, Stitzel JD, Sparks JL, Duma SM (2012) Biomechanical response of human spleen in tensile loading. J Biomech 45:348–355. https://doi.org/10.1016/j.jbiomech.2011.10.022
18. Barnes SC, Shepherd DET, Espino DM, Bryan RT (2015) Frequency dependent viscoelastic properties of porcine bladder. J Mech Behav Biomed Mater 42:168–176. https://doi.org/10.1016/j.jmbbm.2014.11.017
19. Chanda A, Unnikrishnan V, Roy S, Richter HE (2015) Computational modeling of the female pelvic support structures and organs to understand the mechanism of pelvic organ prolapse: a review. Appl Mech Rev 67. https://doi.org/10.1115/1.4030967
20. Ashton-Miller JA, Delancey JOL (2007) Functional anatomy of the female pelvic floor. Ann N Y Acad Sci 1101. https://doi.org/10.1196/annals.1389.034
21. Baah-Dwomoh A, McGuire J, Tan T, De Vita R (2016) Mechanical properties of female reproductive organs and supporting connective tissues: a review of the current state of knowledge. Appl Mech Rev 68. https://doi.org/10.1115/1.4034442
22. Li X, Kruger JA, Chung J-H, Nash MP, Nielsen PMF (2008) Modelling childbirth: comparing athlete and non-athlete pelvic floor mechanics. Springer, Berlin, Heidelberg, pp 750–757. https://doi.org/10.1007/978-3-540-85990-1_90
23. Dhandayuthapani B, Yoshida Y, Maekawa T, Kumar DS (2011) Polymeric scaffolds in tissue engineering application: a review. Int J Polym Sci 2011. https://doi.org/10.1155/2011/290602
24. Mayer J, Karamuk E, Akaike T, Wintermantel E (2000) Matrices for tissue engineering-scaffold structure for a bioartificial liver support system. J Control Release 6481–90 (Elsevier). https://doi.org/10.1016/S0168-3659(99)00136-4
25. Oberpenning F, Meng J, Yoo JJ, Atala A (1999) De novo reconstitution of a functional mammalian urinary bladder by tissue engineering. Nat Biotechnol 17:149–155. https://doi.org/10.1038/6146

26. Tziampazis E, Sambanis A (1995) Tissue engineering of a bioartificial pancreas: modeling the cell environment and device function. Biotechnol Prog 11:115–126. https://doi.org/10.1021/bp00032a001
27. Pabari A, Lloyd-Hughes H, Seifalian AM, Mosahebi A (2014) Nerve conduits for peripheral nerve surgery. Plast Reconstr Surg 133:1420–1430. https://doi.org/10.1097/PRS.0000000000000226
28. Germain L, Auger FA, Grandbois E, Guignard R, Giasson M, Boisjoly H et al (1999) Reconstructed human cornea produced in vitro by tissue engineering. Pathobiology 67:140–147. https://doi.org/10.1159/000028064
29. Bressan E, Favero V, Gardin C, Ferroni L, Iacobellis L, Favero L et al (2011) Biopolymers for hard and soft engineered tissues: application in odontoiatric and plastic surgery field. Polym (Basel) 3:509–526. https://doi.org/10.3390/polym3010509
30. Del Bakhshayesh AR, Asadi N, Alihemmati A, Tayefi Nasrabadi H, Montaseri A, Davaran S et al (2019) An overview of advanced biocompatible and biomimetic materials for creation of replacement structures in the musculoskeletal systems: focusing on cartilage tissue engineering. J Biol Eng 13:1–21. https://doi.org/10.1186/s13036-019-0209-9
31. West JL (2020) Biomaterials for cardiovascular tissue engineering. Biomater Sci 1389–1397 (Elsevier). https://doi.org/10.1016/b978-0-12-816137-1.00086-6
32. Nguyen PK, Baek K, Deng F, Criscione JD, Tuan RS, Kuo CK (2020) Tendon tissue-engineering scaffolds. Biomater Sci 1351–1371 (Elsevier). https://doi.org/10.1016/b978-0-12-816137-1.00084-2
33. Siddiqui N, Asawa S, Birru B, Baadhe R, Rao S (2018) PCL-based composite scaffold matrices for tissue engineering applications. Mol Biotechnol 60:506–532. https://doi.org/10.1007/s12033-018-0084-5
34. Talebian S, Mehrali M, Taebnia N, Pennisi CP, Kadumudi FB, Foroughi J et al (2019) Self-healing hydrogels: the next paradigm shift in tissue engineering? Adv Sci 6. https://doi.org/10.1002/advs.201801664
35. Wang Y, Adokoh CK, Narain R (2018) Recent development and biomedical applications of self-healing hydrogels. Expert Opin Drug Deliv 15:77–91. https://doi.org/10.1080/17425247.2017.1360865
36. Mondal S, Das S, Nandi AK (2020) A review on recent advances in polymer and peptide hydrogels. Soft Matter 16:1404–1454. https://doi.org/10.1039/c9sm02127b
37. Cheng S, Gandevia SC, Green M, Sinkus R, Bilston LE (2011) Viscoelastic properties of the tongue and soft palate using MR elastography. J Biomech 44:450–454. https://doi.org/10.1016/j.jbiomech.2010.09.027
38. Huang YP, Zheng YP, Wang SZ, Chen ZP, Huang QH, He YH (2009) An optical coherence tomography (OCT)-based air jet indentation system for measuring the mechanical properties of soft tissues. Meas Sci Technol 20:015805. https://doi.org/10.1088/0957-0233/20/1/015805
39. Lu MH, Yu W, Huang QH, Huang YP, Zheng YP (2009) A hand-held indentation system for the assessment of mechanical properties of soft tissues in vivo. IEEE Trans Instrum Meas 58:3079–3085. https://doi.org/10.1109/TIM.2009.2016876
40. Green MA, Bilston LE, Sinkus R (2008) In vivo brain viscoelastic properties measured by magnetic resonance elastography. NMR Biomed 21:755–764. https://doi.org/10.1002/nbm.1254
41. Sack I, Rump J, Elgeti T, Samani A, Braun J (2009) MR elastography of the human heart: noninvasive assessment of myocardial elasticity changes by shear wave amplitude variations. Magn Reson Med 61:668–677. https://doi.org/10.1002/mrm.21878
42. Huwart L, Peeters F, Sinkus R, Annet L, Salameh N, ter Beek LC et al (2006) Liver fibrosis: non-invasive assessment with MR elastography. NMR Biomed 19:173–179. https://doi.org/10.1002/nbm.1030
43. Lorenzen J, Sinkus R, Biesterfeldt M, Adam G (2003) Menstrual-cycle dependence of breast parenchyma elasticity: estimation with magnetic resonance elastography of breast tissue during the menstrual cycle. Invest Radiol 38:236–240. https://doi.org/10.1097/01.rli.0000059544.18910.bd

44. Uffmann K, Maderwald S, Ajaj W, Galban CG, Mateiescu S, Quick HH et al (2004) In vivo elasticity measurements of extremity skeletal muscle with MR elastography. NMR Biomed 17:181–190. https://doi.org/10.1002/nbm.887
45. Domire ZJ, McCullough MB, Chen Q, An KN (2009) Feasibility of using magnetic resonance elastography to study the effect of aging on shear modulus of skeletal muscle. J Appl Biomech 25:93–97. https://doi.org/10.1123/jab.25.1.93
46. Ottensmeyer MP, Kerdok AE, Howe RD, Dawson SL (2004) The effects of testing environment on the viscoelastic properties of soft tissues. Lect Notes Comput Sci (Including Subser Lect Notes Artif Intell Lect Notes Bioinformatics) 3078:9–18. https://doi.org/10.1007/978-3-540-25968-8_2
47. Scholze M, Safavi S, Li KC, Ondruschka B, Werner M, Zwirner J et al (2020) Standardized tensile testing of soft tissue using a 3D printed clamping system. HardwareX 8:e00159. https://doi.org/10.1016/J.OHX.2020.E00159
48. Cooney GM, Moerman KM, Takaza M, Winter DC, Simms CK (2015) Uniaxial and biaxial mechanical properties of porcine linea alba. J Mech Behav Biomed Mater 41:68–82. https://doi.org/10.1016/J.JMBBM.2014.09.026
49. Chow MJ, Zhang Y (2011) Changes in the mechanical and biochemical properties of aortic tissue due to cold storage. J Surg Res 171:434–442. https://doi.org/10.1016/J.JSS.2010.04.007
50. Jacobs NT, Cortes DH, Vresilovic EJ, Elliott DM (2013) Biaxial tension of fibrous tissue: Using finite element methods to address experimental challenges arising from boundary conditions and anisotropy. J Biomech Eng 135:1–10. https://doi.org/10.1115/1.4023503/371316
51. Jokandan MS, Ajalloueian F, Edinger M, Stubbe PR, Baldursdottir S, Chronakis IS (2018) Bladder wall biomechanics: a comprehensive study on fresh porcine urinary bladder. J Mech Behav Biomed Mater 79:92–103. https://doi.org/10.1016/j.jmbbm.2017.11.034
52. Griffin M, Premakumar Y, Seifalian A, Butler PE, Szarko M (2016) Biomechanical characterization of human soft tissues using indentation and tensile testing. Journal Vis Exp JoVE e54872. https://doi.org/10.3791/54872
53. Walraevens J, Willaert B, De Win G, Ranftl A, De Schutter J, Vander SJ (2008) Correlation between compression, tensile and tearing tests on healthy and calcified aortic tissues. Med Eng Phys 30:1098–1104. https://doi.org/10.1016/J.MEDENGPHY.2008.01.006
54. Wu JZ, Cutlip RG, Andrew ME, Dong RG (2007) Simultaneous determination of the nonlinear-elastic properties of skin and subcutaneous tissue in unconfined compression tests. Ski Res Technol 13:34–42. https://doi.org/10.1111/J.1600-0846.2007.00182.X
55. Rashid B, Destrade M, Gilchrist MD (2012) Determination of friction coefficient in unconfined compression of brain tissue. J Mech Behav Biomed Mater 14:163–171. https://doi.org/10.1016/J.JMBBM.2012.05.001
56. Haddad SMH, Dhaliwal SS, Rotenberg BW, Ladak HM, Samani A (2020) Estimation of the hyperelastic parameters of fresh human oropharyngeal soft tissues using indentation testing. J Mech Behav Biomed Mater 108:103798. https://doi.org/10.1016/j.jmbbm.2020.103798
57. Jijun S, Haitian Z, Tongtong G (2009) The study of mechanical properties on soft tissue of human forearm in vivo. In: 3rd international conference on bioinformatics and biomedical engineering (iCBBE 2009). https://doi.org/10.1109/ICBBE.2009.5163671
58. Joodaki H, Panzer MB (2018) Skin mechanical properties and modeling: a review. Proc Inst Mech Eng Part H J Eng Med 232:323–343. https://doi.org/10.1177/0954411918759801
59. Sun W, Sacks MS, Scott MJ (2005) Effects of boundary conditions on the estimation of the planar biaxial mechanical properties of soft tissues. J Biomech Eng 127:709–715 (American Society of Mechanical Engineers Digital Collection). https://doi.org/10.1115/1.1933931.

Chapter 2
Skin

2.1 Introduction

The largest organ in terms of surface area, the skin serves a number of purposes, including defence against the external environment, control of body temperature, and adaptability to the body contours during movement. This tissue is mechanically complex since it is made up of several layers, each of which is composed of various components. For many medical, cosmetic, and biomechanical purposes, the skin's mechanical properties play a significant role. Applications in cosmetics include evaluating the emolliency and hydration of cosmetic products. Clinicians make use of the mechanical properties of skin to better comprehend the suturing techniques, skin disease therapies, plastic surgery procedures, etc. [1, 2]. The biomechanical properties of the skin tissues are also crucial for the design and fabrication of skin-simulant materials for real-time anthropomorphic devices like crash test dummies and surgical simulators. Skin material properties are also important employed in the designing and analyzing of computational human body models for biomechanical simulation, in addition to their application with physical biomechanical models [3].

2.2 Structure of Skin

The human skin is an anisotropic material with non-homogeneous and viscoelastic properties and has a thickness variations ranging from 0.5 mm for eyelids (folding and unfolding) to 6 mm for the foot sole [4–6]. In adults, the skin's surface area is 2 m^2, comprising an average body mass of 5.5% [7–9]. The epidermis, dermis, and hypodermis are the three primary layers that make up the complex, multilayered (Fig. 2.1) [10, 11]. The thickness of human skin ranges between 0.3 and 2.6 mm at various body sites [12] and has an average thickness of 2 mm [9]. The dermis layer is mostly made up of fibrous proteins including collagen, reticulin, and elastin, while

A. Chanda and G. Singh, *Mechanical Properties of Human Tissues*,
Materials Horizons: From Nature to Nanomaterials,
https://doi.org/10.1007/978-981-99-2225-3_2

the epidermis layer is a thin layer that covers the majority of the human body and is between 0.07 and 0.12 mm thick, measuring 0.8 mm on the palms and 1.4 mm on the soles. Collagen fibers, which range in size from 60 to 100 nm, get denser and thicker as skin depth rises [13]. These collagen fibers, which make up around 75% of the dry weight of the dermal tissue, are responsible for the strength of human skin [14]. Dermis thickness varies from an average of 1–2 mm to as little as 0.6 mm in certain areas [15], and it is widely recognized as a key factor in the mechanical behavior of skin tissue [11]. The hypodermis, which lies under the dermis layer and is closely connected without having a defined border, is mostly composed of connective tissue and fat lobules [16–18]. The hypodermis layer, which is the third layer of skin, is linked to the fascia by a honeycomb structure, and it also serves as the body's fat storage area with a close connection to the dermis and a flexible loose connection with other tissues (such as skin-internal soft tissue).

However, depending on the frictional forces, the epidermis thickness varies throughout the body and affects the stress–strain responses of the human skin (Fig. 2.2). The stratum corneum, stratum granulosum, stratum spinosum, and stratum spinosum are further sublayers that make up the epidermis layer. The dermis, which is below the epidermis layer and has an elastic support structure, is known to have a crucial role in the mechanical behavior of the skin tissue [11]. It is primarily composed of nerve endings, blood capillaries, hair follicles, sweat glands, and lymphatics. The dermis thickness is measured as 0.7 mm on the eyelids and 1.5 mm on the back, belly, buttock, thigh, and dorsum of the foot [15, 20]. The dermis varies in thickness throughout the body (average 1–2 mm) as well. Keratinocytes, cells, and

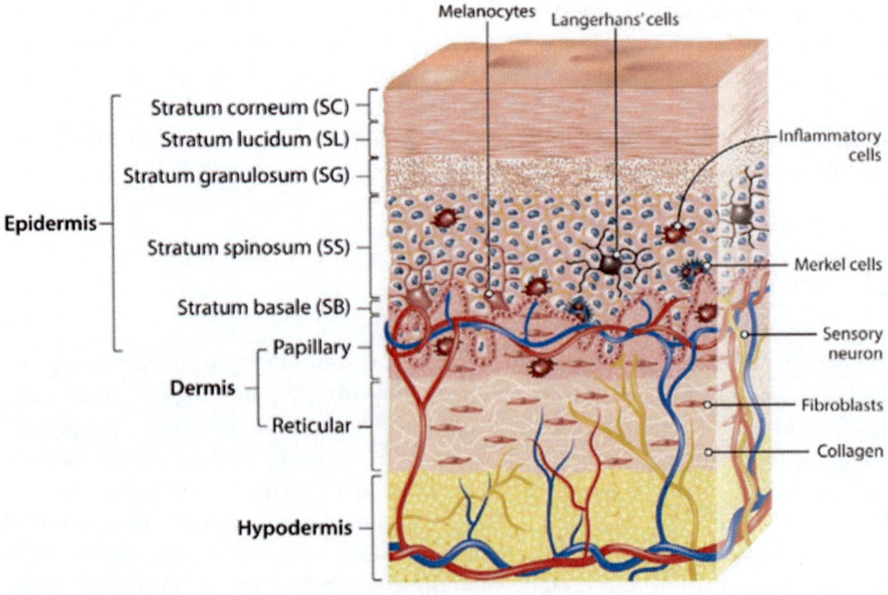

Fig. 2.1 Multilayer structure of human skin tissue [19]

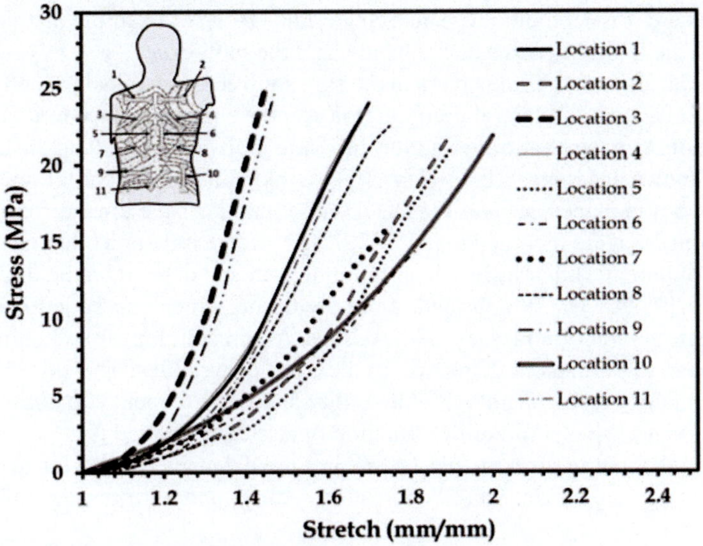

Fig. 2.2 Stress–stretch plots of human skin test data [22]

cellular debris are found in the epidermis layer, while fibroblasts and fibrous proteins including elastin, collagen fibers, and reticulin are the main components of the dermis layer. The collagen fibers are made up of collagen fibrils, which are dense, thick fibers that range in size from 60 to 100 nm [13]. The papillary dermis (thin upper layer) and reticular dermis (thick lower layer), which are located deep to the epidermal layer and superficial to the subcutaneous layer, are the connective tissues of the dermis layer. Collagen fibers, which make up roughly 75% of the dermal tissue's dry weight [14] and elastin fibers, which make up 2–4% of the dry skin weight [21], are what give human skin its strength.

2.3 Mechanical Properties of Skin

Due to variations in dermal thickness and the quantity of collagen and elastin, the mechanical properties of the skin are correlated with skin thickness [23]. The hypodermis, often known as subcutaneous fat, is the third layer of the human skin. It is tightly linked to the dermis layer without having a firm border and contains fat lobules, bigger blood vessels, nerves, and connective tissue [16–18]. The hypodermis layer works as energy storage, maintaining body heat, and serving as a shock absorber to lessen the effects of damage on the tissues and organs below. The hypodermis layer offers a flexible connection with other tissues and is joined to the fascia via a honeycomb structure (i.e., skin-internal soft tissue). The hypodermis thickness also varies based on where it is on the body. The hypodermis layer, for example, is thickest on the soles of the feet, the palms of the hands, and the buttocks.

Anisotropic thin membrane layers in the skin showed mechanical properties such resistance to friction and lateral compression response [24–26]. Due to its irregular shape and nonlinear stress–strain relationship, skin is always under tension. Prestress lines, also known as Langer's lines (Fig. 2.3), wrinkle lines, and contour lines (interaction of two skin planes, such as the cheek–nose junction), are three different kinds of natural lines that are included in it [27–30]. The stretch ratio and other mechanical properties of human skin are significantly influenced and depend on the direction of Langer lines [9, 24]. The skin tissue is a passive tissue, which consequently deforms with the deformation of underneath soft tissues. As a result, longer stretching of the skin may lead to permanent distortion in the form of wrinkles. The skin, however, extends reversibly with considerable flexibility and has a tendency to come back to its previous state, when a minor deformation or stretch is applied [9].

Both, in vivo and in vitro testing techniques are extensively used for the biomechanical testing of the skin. Under controlled settings and with fewer confounding

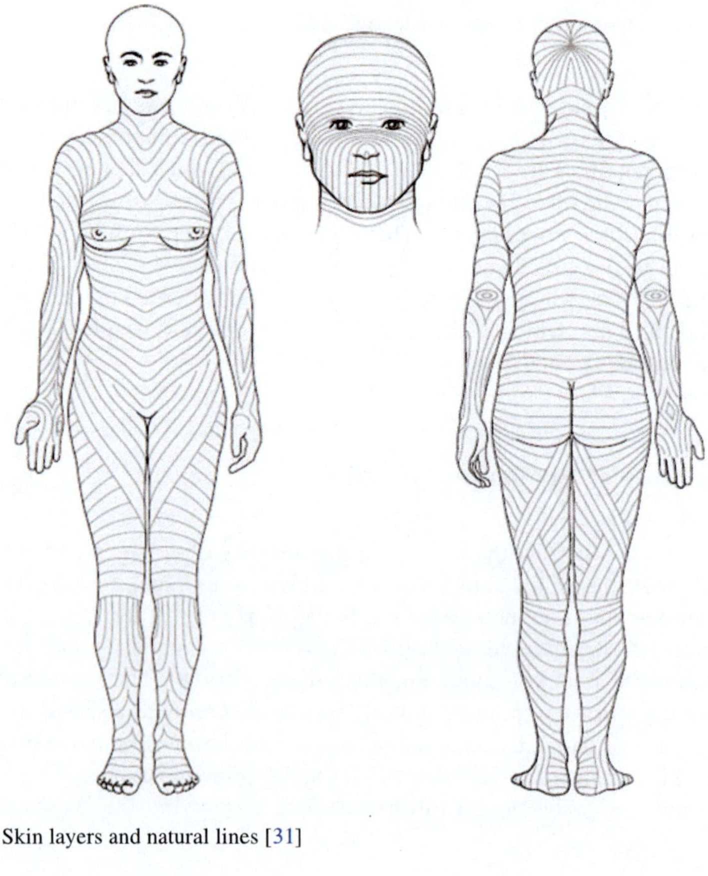

Fig. 2.3 Skin layers and natural lines [31]

variables, in vitro testing provides the stress–strain responses of the tested sample, which can be further investigated to evaluate its mechanical behavior. The ultimate tensile stress and strain at the point of skin rupture may also be determined via in vitro testing. However, because the excised skin is no longer linked to the body, it might be challenging to clamp samples without exerting an axial stress, and its structural integrity is affected, especially at the sample's edges [24]. In contrast, in vivo tensile measurements may take into account the impact of anatomical and physiological changes on skin characteristics. For instance, skin ageing has a detrimental effect on the skin's capacity to regulate body temperature and prevent water loss. Therefore, in vivo longitudinal investigations of skin's Young's modulus values are required.

For the in vivo testing of skin samples, indentation, torsion, and tension are three widely used techniques to estimate its mechanical properties. Indentation investigations include applying a known deformation force to the skin using a rigid indenter. The applied indenter equipped with a cylindrical, conical, or spherical indentation tip, which spreads the stress over the area and estimates its mechanical properties. In torsion testing, the skin is often subjected to a steady rotation or torque via a disk. The coil moved until an equilibrium position was attained as a result of the opposing moment caused by the skin's rotation [32]. For the tensile test, the sample of skin loaded parallelly to its surface and attached using the adhesive materials (e.g., double-sided tape) to avoid the creep deformation. These three are the basic methods to test the skin samples. However, the most widely used are the uniaxial tension and biaxial tension, which have numerous advantages compared to the in vivo testing techniques. The ability to test the tissue sample under varying loading conditions, strain rates, and easy evaluation of the stress–strain curve is one of prominence of these methods. Another benefit is that failure properties can be identified since the tissue can be evaluated all the way to failure. Furthermore, as the boundary conditions are well stated, it offers stress–strain relationships that are simple to model and quantify. Along with several advantages, both uniaxial and biaxial have some limitations. First, between the excision and the testing period, the sample is often stored, which may change the mechanical characteristics of the tissue. Second, the sample is difficult to handle and retain since the soft tissue has a tendency to slip from the clamp, during the testing. For uniaxial tensile testing, the dog-bone-shaped sliced skin sample is kept fixed from one end and pulled from the other end. To ascertain the material's anisotropic behavior, multiple tests in different directions are required to estimate the accurate mechanical properties of the tested sample. The fixation methods such employing a freeze clamp, a serrated jaw clamp, or wrapping the tissue in sandpaper paper would be beneficial to avoid the sample slippage during the testing [33–36]. The uniaxial tensile testing is the only one that can be used to evaluate the failure characteristics of a soft tissue. Under biaxial testing conditions, the tissue samples (in square shape) are attached using strings and hooks to avoid shear strains, and load is applied biaxially to stretch the sample in both the directions. However, biaxial testing may be difficult since the hooks attached to the sample's edges have a tendency to rip the skin. The failure properties of the tested sample cannot be measured using this approach for the same reason.

The main load-bearer in the skin's outer layer, the stratum corneum, serves as a barrier between the environment and the body (i.e., epidermal layer). It exhibits nonlinear, isotropic, stiff behavior with a Poisson's ratio of 0.49, and the degree of hydration and temperature has a substantial impact on the mechanical properties of skin [37, 38]. According to Wu et al. [39], the human stratum corneum's young's modulus ranges from 1000 to 5 MPa with respect to hydration variations between 0 and 100%. However, for dry and thoroughly hydrated stratum corneum, Yuan and Verma [40] indicated 200 and 50 MPa elasticity modulus. The epidermal skin layer's biomechanical characteristics ultimately convey stresses and strains to the mechanoreceptors, producing tactile feedback that generates sensation. The temperature factor has a considerable impact on the mechanical characteristics of human skin. Due to the breakage of molecular cross-links at high temperatures, the skin's collagen fibers transform into a gel-like substance, reducing stiffness during the strain hardening phase of the tensile test [9]. Annaidh et al. [24] revealed the tensile strength of human skin tissue, with a mean failure strain of 54% ± 17%. They showed that the average ultimate tensile strength was 21.6 ± 8.4 MPa and the elastic modulus was 83.3 ± 34.9 MPa, respectively. Edwards and Marks [7] reported the range of tensile strength from 5 to 30 MPa, with the mean value at 8 years being 21 MPa and a lower mean value of 17 MPa being achieved at the 95 years of age. Similarly, the modulus of elasticity ranges from 15 to 150 MPa, with the emphasized values for the chosen age groups being 70 MPa and 60 MPa, respectively. According to Ottenio et al. [9], who tested the human skin under uniaxial tensile conditions, the mean elastic modulus and mean tensile stress were directly proportional to the strain rates, i.e., it increases as the strain rate increases. In samples oriented parallel to the Langer lines, the measured values of ultimate tensile stress and elastic modulus were 28 ± 5.7 MPa and 160.8 ± 53.2 MPa, respectively. While the elastic modulus and ultimate tensile stress for perpendicular orientation were 70.6 ± 59.5 MPa and 15.6 ± 5.2 MPa, respectively.

2.4 Skin Friction

Over the years, the tribology of human skin has been a topic of great attention. Usually, tribological skin research focused on cosmetics and their effects on the human skin [41, 42]. Other research areas may include skin ageing, skin injuries, prosthetics, wound healing, medical and sports applications and automotive applications [43–45]. The surface and friction properties are important to evaluate the tactile properties of materials and surfaces. Due to these factors, attempts have been made for making mechanical skin models with tribological testing, which can replicate the realistic mechanical interactions between the skin and external surfaces [46, 47]. The material and surface characteristics of the skin, the contacting material, as well as potential intermediate layers like temporarily trapped or topically applied substances (like cosmetics) or sweat and sebum naturally excreted from skin into the tribointerface, determine the friction coefficient of human skin. Skin friction depends on the type (solid, soft, and fibrous material) and physical characteristics of the materials contacting the skin, as well as on physiological skin conditions (such as hydration level and sebum level), and mechanical contact parameters, especially on the normal load. According to Dowson [48], the coefficient of friction increases with dry skin because of adhesion brought on by attractive surface forces at the skin–material interface and deformation (hysteresis, ploughing) of the softer and viscoelastic bulk skin tissue. Human skin friction is considered to be primarily caused by adhesion, with deformation processes contributing just slightly [49]. The mechanical contact behavior and friction processes of skin have been discussed and described in the literature using a variety of theoretical models, including those by Hertz, Johnson–Kendall–Roberts, and Greenwood–Williamson [50, 51]. It should be noted that the load dependence of recorded friction coefficients varies across different friction mechanisms [52].

Depending on the location of the body, the human skin's surface topography can be distinguished by either concentric ridges (found on finger pads) or furrows (found on, say, the forearm) that define polygonal sections of varying sizes. The typical surface roughness values Ra and Rz, which correspond to the relief of first order furrows (70–200 μm) and that of second order furrows (20–70 μm), respectively, are in the ranges of (10–30 μm and 30–140 μm). The skin anisotropy, furrow spacing, and skin roughness have been affected by the aging and increases with the age [53, 54]. The skin's surface is typically shielded by an acidic hydrolipid coating (pH 4–6), which regulates skin flora, prevents pathogenic species from colonizing the skin, and serves as defence against invasive microbes. The hydrolipid film, which coats the stratum corneum as a water–oil emulsion, is made of sebum from sebaceous glands and sebum from sweat glands [55, 56]. According to Pailler-Mattei et al. [57], capillary processes cause the skin surface lipid coating to affect the skin adhesion qualities. While a considerable adhesion force could be measured on normal skin, once the lipid layer was removed, the adhesion force decreased. Cua et al. [55] noted modest associations, particularly on the forehead ($r = 0.33$) and postauricular skin ($r = 0.41$) between the lipid content of the skin surface and friction. Moisture often makes the skin surface more abrasive, as seen in daily life, for example, when perspiration causes a piece of clothing to cling to the skin during physical activity. Skin friction coefficients have been shown to differ between wet and dry circumstances by factors of 1.5–7 [58, 59]. This wide variation in test procedures, materials, and experimental settings is probably reason for such issues. The time interval between a friction measurement and the application of moisturizer or skin exposure to water is one of the most significant aspects. Recent studies specifically reported at the functional and qualitative link between friction and skin moisture. Skin moisture and friction have been shown to exhibit bell-shaped, exponential, power-law, linear, and exponential correlations [60–63].

According to a recent study by Tomlinson et al. [64], investigations on grip characteristics, or feel elements of engineering surfaces, are the source of a large portion of today's information concerning the impact of a contacting material's surface roughness on skin friction. The COF between a hard surface and naturally dry skin (fingertip or hand) often decreases with increasing material surface roughness [61] when Ra changed in the range of 0.03–11.5 μm, Rz = 0.05–45 μm [65]. Masen et al. [65] evaluated the roughness and coefficient of friction for the hydrated skin and reported the roughness values as Rq = 0.004–2 μm and 0.9–1.7 for friction coefficient, respectively. They revealed that the adhesion and deformation elements in the state of hydrated skin were causing the high friction in the intermediate roughness regime. The majority of research estimating the skin friction at various anatomical regions revealed no significant variations in gender or age [66]. At various anatomical areas, previous studies have shown significant variations in the skin's frictional qualities with a tendency of high friction at high skin hydration. It was reported that the friction on the finger pad, hand's palm, forehead, and vulva was greater than that on the hand's edge, belly, thighs, legs, and lower back [67, 68].

References

1. Diridollou S, Patat F, Gens F, Vaillant L, Black D, Lagarde JM et al (2000) In vivo model of the mechanical properties of the human skin under suction. Ski Res Technol 6:214–221. https://doi.org/10.1034/J.1600-0846.2000.006004214.X
2. Hochberg J, Meyer KM, Marion MD (2009) Suture choice and other methods of skin closure. Surg Clin North Am 89:627–641. https://doi.org/10.1016/J.SUC.2009.03.001
3. Untaroiu CD, Yue N, Shin J (2013) A finite element model of the lower limb for simulating automotive impacts. Ann Biomed Eng 41:513–526. https://doi.org/10.1007/S10439-012-0687-0/FIGURES/13
4. Crichton ML, Donose BC, Chen X, Raphael AP, Huang H, Kendall MAF (2011) The viscoelastic, hyperelastic and scale dependent behaviour of freshly excised individual skin layers. Biomaterials 32:4670–4681. https://doi.org/10.1016/j.biomaterials.2011.03.012
5. Lapeer RJ, Gasson PD, Karri V (2010) Simulating plastic surgery: from human skin tensile tests, through hyperelastic finite element models to real-time haptics. Prog Biophys Mol Biol 103:208–216. https://doi.org/10.1016/j.pbiomolbio.2010.09.013
6. Kumaraswamy N, Khatam H, Reece GP, Fingeret MC, Markey MK, Ravi-Chandar K (2017) Mechanical response of human female breast skin under uniaxial stretching. J Mech Behav Biomed Mater 74:164–175. https://doi.org/10.1016/j.jmbbm.2017.05.027
7. Edwards C, Marks R (1995) Evaluation of biomechanical properties of human skin. Clin Dermatol 13:375–380. https://doi.org/10.1016/0738-081X(95)00078-T
8. Goldsmith LA (1990) My organ is bigger than your organ. Arch Dermatol 126:301–302. https://doi.org/10.1001/archderm.1990.01670270033005
9. Ottenio M, Tran D, Ní Annaidh A, Gilchrist MD, Bruyère K (2015) Strain rate and anisotropy effects on the tensile failure characteristics of human skin. J Mech Behav Biomed Mater 41:241–250. https://doi.org/10.1016/j.jmbbm.2014.10.006
10. Chanda A, Graeter R, Unnikrishnan V (2015) Effect of blasts on subject-specific computational models of skin and bone sections at various locations on the human body. AIMS Mater Sci 2:425–447. https://doi.org/10.3934/matersci.2015.4.425
11. Groves RB, Coulman SA, Birchall JC, Evans SL (2013) An anisotropic, hyperelastic model for skin: experimental measurements, finite element modelling and identification of parameters for human and murine skin. J Mech Behav Biomed Mater 18:167–180. https://doi.org/10.1016/j.jmbbm.2012.10.021
12. Fenton LA, Horsfall I, Carr DJ (2020) Skin and skin simulants. Aust J Forensic Sci 52:96–106. https://doi.org/10.1080/00450618.2018.1450896
13. Silver FH, Freeman JW, Devore D (2001) Viscoelastic properties of human skin and processed dermis. Ski Res Technol 7:18–23. https://doi.org/10.1034/j.1600-0846.2001.007001018.x

14. Wilkes GL, Brown IA, Wildnauer RH (1973) The biomechanical properties of skin. CRC Crit Rev Bioeng 1:453–495
15. Pailler-Mattei C, Bec S, Zahouani H (2008) In vivo measurements of the elastic mechanical properties of human skin by indentation tests. Med Eng Phys 30:599–606. https://doi.org/10.1016/j.medengphy.2007.06.011
16. Geerligs M, Peters GWM, Ackermans PAJ, Oomens CWJ, Baaijens FPT (2008) Linear viscoelastic behavior of subcutaneous adipose tissue. Biorheology 45:677–688. https://doi.org/10.3233/BIR-2008-0517
17. Miller-Young JE, Duncan NA, Baroud G (2002) Material properties of the human calcaneal fat pad in compression: experiment and theory. J Biomech 35:1523–1531. https://doi.org/10.1016/S0021-9290(02)00090-8
18. McGrath JA, Uitto J (2010) Anatomy and organization of human skin. Rook's Textb Dermatol 1:1–53 (Oxford, UK: Wiley-Blackwell; 2010). https://doi.org/10.1002/9781444317633.ch3
19. Rahmati M, Blaker JJ, Lyngstadaas SP, Mano JF, Haugen HJ (2020) Designing multigradient biomaterials for skin regeneration. Mater Today Adv 5:100051. https://doi.org/10.1016/J.MTADV.2019.100051
20. Lee Y, Hwang K (2002) Skin thickness of Korean adults. Surg Radiol Anat 24:183–189. https://doi.org/10.1007/s00276-002-0034-5
21. Pissarenko A, Yang W, Quan H, Brown KA, Williams A, Proud WG et al (2019) Tensile behavior and structural characterization of pig dermis. Acta Biomater 86:77–95. https://doi.org/10.1016/j.actbio.2019.01.023
22. Chanda A (2018) Biomechanical modeling of human skin tissue surrogates. Biomimetics 3:18. https://doi.org/10.3390/BIOMIMETICS3030018
23. Sutradhar A, Miller MJ (2013) In vivo measurement of breast skin elasticity and breast skin thickness. Ski Res Technol 19:e191–e199. https://doi.org/10.1111/j.1600-0846.2012.00627.x
24. Ní Annaidh A, Bruyère K, Destrade M, Gilchrist MD, Otténio M (2012) Characterization of the anisotropic mechanical properties of excised human skin. J Mech Behav Biomed Mater 5:139–148. https://doi.org/10.1016/j.jmbbm.2011.08.016
25. Wu JZ, Dong RG, Smutz WP, Schopper AW (2003) Nonlinear and viscoelastic characteristics of skin under compression: experiment and analysis. Biomed Mater Eng 13:373–385
26. Flynn C, McCormack BAO (2008) Finite element modelling of forearm skin wrinkling. Ski Res Technol 14:261–269. https://doi.org/10.1111/j.1600-0846.2008.00289.x
27. Langer K (1978) On the anatomy and physiology of the skin. I. The cleavability of the cutis. Br J Plast Surg 31:3–8 (Churchill Livingstone). https://doi.org/10.1016/0007-1226(78)90003-6
28. Kwak M, Son D, Kim J, Han K (2014) Static Langer's line and wound contraction rates according to anatomical regions in a porcine model. Wound Repair Regen 22:678–682. https://doi.org/10.1111/wrr.12206
29. Cox HT (1941) The cleavage lines of the skin. Br J Surg 29:234–240. https://doi.org/10.1002/bjs.18002911408
30. Pissarenko A, Meyers MA (2020) The materials science of skin: analysis, characterization, and modeling. Prog Mater Sci 110:100634. https://doi.org/10.1016/j.pmatsci.2019.100634
31. Newell KA (2007) Wound closure. Essent Clin Proced 313–341. https://doi.org/10.1016/B978-1-4160-3001-0.50027-7
32. Sanders R (1973) Torsional elasticity of human skin in vivo. Pflügers Arch Eur J Physiol 342:255–260. https://doi.org/10.1007/BF00591373/METRICS
33. Karimi A, Faturechi R, Navidbakhsh M, Hashemi SA (2014) A Nonlinear hyperelastic behavior to identify the mechanical properties of rat skin under uniaxial loading. J Mech Med Biol 14. https://doi.org/101142/S0219519414500754
34. Karimi A, Navidbakhsh M (2015) Measurement of the uniaxial mechanical properties of rat skin using different stress–strain definitions. Ski Res Technol 21:149–157. https://doi.org/10.1111/SRT.12171
35. Velardi F, Fraternali F, Angelillo M (2006) Anisotropic constitutive equations and experimental tensile behavior of brain tissue. Biomech Model Mechanobiol 5:53–61. https://doi.org/10.1007/S10237-005-0007-9/METRICS

36. Shi DF, Wang DM, Wang CT, Liu A (2012) A novel, inexpensive and easy to use tendon clamp for in vitro biomechanical testing. Med Eng Phys 34:516–520. https://doi.org/10.1016/J.MEDENGPHY.2011.11.019
37. Chanda A, Unnikrishnan V (2017) A realistic 3D computational model of the closure of skin wound with interrupted sutures. J Mech Med Biol 17. https://doi.org/10.1142/S0219519417500257
38. Chanda A, Graeter R (2018) Human skin-like composite materials for blast induced injury mitigation. J Compos Sci 2:44. https://doi.org/10.3390/jcs2030044
39. Wu KS, Van Osdol WW, Dauskardt RH (2006) Mechanical properties of human stratum corneum: effects of temperature, hydration, and chemical treatment. Biomaterials 27:785–795. https://doi.org/10.1016/j.biomaterials.2005.06.019
40. Yuan Y, Verma R (2006) Measuring microelastic properties of stratum corneum. Colloids Surf B Biointerfaces 48:6–12. https://doi.org/10.1016/j.colsurfb.2005.12.013
41. Sivamani RK, Wu GC, Gitis NV, Maibach HI (2003) Tribological testing of skin products: gender, age, and ethnicity on the volar forearm. Ski Res Technol 9:299–305. https://doi.org/10.1034/J.1600-0846.2003.00034.X
42. Tang W, Bhushan B, Ge S (2010) Friction, adhesion and durability and influence of humidity on adhesion and surface charging of skin and various skin creams using atomic force microscopy. J Microsc 239:99–116. https://doi.org/10.1111/J.1365-2818.2009.03362.X
43. Cua AB, Wilhelm KP, Maibach HI (1990) Frictional properties of human skin: relation to age, sex and anatomical region, stratum corneum hydration and transepidermal water loss. Br J Dermatol 123:473–479. https://doi.org/10.1111/J.1365-2133.1990.TB01452.X
44. Lodén M, Olsson H, Axéll T, Linde YW (1992) Friction, capacitance and transepidermal water loss (TEWL) in dry atopic and normal skin. Br J Dermatol 126:137–141. https://doi.org/10.1111/J.1365-2133.1992.TB07810.X
45. Liu X, Chan MK, Hennessey B, Rübenach T, Alay G (2005) Quantifying touch-feel perception on automotive interiors by a multi-function tribological probe microscope. J Phys Conf Ser 13:357. https://doi.org/10.1088/1742-6596/13/1/082
46. Bhushan B, Wei G, Haddad P (2005) Friction and wear studies of human hair and skin. Wear 259:1012–1021. https://doi.org/10.1016/J.WEAR.2004.12.026
47. Derler S, Schrade U, Gerhardt LC (2007) Tribology of human skin and mechanical skin equivalents in contact with textiles. Wear 263:1112–1116. https://doi.org/10.1016/J.WEAR.2006.11.031
48. Wilhelm K-P, Elsner P, Berardesca E, Maibach HI (2007) Bioengineering of the skin: Skin imaging & analysis (2nd ed). CRC press. https://doi.org/10.3109/9781420005516
49. Adams MJ, Briscoe BJ, Johnson SA (2007) Friction and lubrication of human skin. Tribol Lett 26:239–253. https://doi.org/10.1007/S11249-007-9206-0/FIGURES/13
50. Koudine AA, Barquins M, Anthoine PH, Aubert L, Leveque JL (2000) Frictional properties of skin: proposal of a new approach. Int J Cosmet Sci 22:11–20. https://doi.org/10.1046/J.1467-2494.2000.00006.X
51. Pailler-Mattéi C, Zahouani H (2012) Study of adhesion forces and mechanical properties of human skin in vivo. Br J Dermatol 18:1739–58. https://doi.org/10.1163/1568561042708368.
52. Derler S, Gerhardt LC, Lenz A, Bertaux E, Hadad M (2009) Friction of human skin against smooth and rough glass as a function of the contact pressure. Tribol Int 42:1565–1574. https://doi.org/10.1016/J.TRIBOINT.2008.11.009
53. Akazaki S, Nakagawa H, Kazama H, Osanai O, Kawai M, Takema Y et al (2002) Age-related changes in skin wrinkles assessed by a novel three-dimensional morphometric analysis. Br J Dermatol 147:689–695. https://doi.org/10.1046/J.1365-2133.2002.04874.X
54. Boyer G, Laquièze L, Le Bot A, Laquièze S, Zahouani H (2009) Dynamic indentation on human skin in vivo: ageing effects. Ski Res Technol 15:55–67. https://doi.org/10.1111/J.1600-0846.2008.00324.X
55. Cua AB, Wilhelm KP, Maibach HI (1995) Skin surface lipid and skin friction: relation to age, sex and anatomical region. Skin Pharmacol Physiol 8:246–251. https://doi.org/10.1159/000211354

56. El Khyat A, Mavon A, Leduc M, Agache P, Humbert P (1996) Skin critical surface tension. Ski Res Technol 2:91–96. https://doi.org/10.1111/J.1600-0846.1996.TB00066.X
57. Pailler-Mattei C, Nicoli S, Pirot F, Vargiolu R, Zahouani H (2009) A new approach to describe the skin surface physical properties in vivo. Colloids Surfaces B Biointerfaces 68:200–206. https://doi.org/10.1016/J.COLSURFB.2008.10.005
58. Comaish S, Bottoms E (1971) The skin and friction: deviations from Amonton's laws, and the effects of hydration and lubrication. Br J Dermatol 84:37–43. https://doi.org/10.1111/J.1365-2133.1971.TB14194.X
59. Pavlov Y V, Rae A, Bruce R, Stout HP, Townend PP, Matto AR et al (2016) Influence of fiber type and moisture on measured fabric-to-skin friction. Textile Res J 64:722–728. https://doi.org/10.1177/004051759406401204
60. Gerhardt LC, Strässle V, Lenz A, Spencer ND, Derler S (2008) Influence of epidermal hydration on the friction of human skin against textiles. J R Soc Interface 5:1317–1328. https://doi.org/10.1098/RSIF.2008.0034
61. Hendriks CP, Franklin SE (2010) Influence of surface roughness, material and climate conditions on the friction of human skin. Tribol Lett 37:361–373. https://doi.org/10.1007/S11249-009-9530-7/TABLES/3
62. Kwiatkowska M, Franklin SE, Hendriks CP, Kwiatkowski K (2009) Friction and deformation behaviour of human skin. Wear 267:1264–1273. https://doi.org/10.1016/J.WEAR.2008.12.030
63. André T, Lefèvre P, Thonnard JL (2009) A continuous measure of fingertip friction during precision grip. J Neurosci Methods 179:224–229. https://doi.org/10.1016/J.JNEUMETH.2009.01.031
64. Tomlinson SE, Lewis R, Carré MJ (2007) Review of the frictional properties of finger-object contact when gripping. Proc Inst Mech Eng Part J J Eng Tribol 221:841–850. https://doi.org/10.1243/13506501JET313
65. Masen MA (2011) A systems based experimental approach to tactile friction. J Mech Behav Biomed Mater 4:1620–1626. https://doi.org/10.1016/J.JMBBM.2011.04.007
66. O'Meara DM, Smith RM (2010) Static friction properties between human palmar skin and five grabrail materials. Ergonomics 44:973–88. https://doi.org/10.1080/00140130110074882
67. Zhang M, Mak AFT (2016) In vivo friction properties of human skin. http://doi.org/103109/03093649909071625. 23:135–141. https://doi.org/10.3109/03093649909071625
68. Elsner P, Wilhelm D, Maibach HI (1990) Physiological skin surface water loss dynamics of human vulvar and forearm skin. Acta Derm Venereol 70:141–144. https://doi.org/102340/0001555570141144

Chapter 3
Muscles and Connective Tissues

3.1 Muscles

Muscles are the flexible tissues with contractile properties and are viscoelastic, anisotropic, and nonlinear in nature. Muscle tissues are responsible for controlling body parts or organs to perform different functions [1]. The muscle tissue can be categorized as smooth muscle, skeletal muscle, and cardiac muscle (Fig. 3.1). Smooth muscle functions without your conscious effort (involuntary muscle). It may be found in the blood artery walls and the walls of hollow bodily organs such the stomach, intestines, bladder, and uterus. Smooth muscle enables organs to loosen up and expand or tighten up and constrict (contract). Skeletal muscles are the voluntary muscles and control the body to move it. Skeletal muscles hold the bones together and are connected to the bones directly. Additionally, these muscles enable the human body to move different body parts, including your arms and legs. The heart's walls are made of cardiac muscle, which also enables the heart to pump blood. Cardiac muscle functions autonomously and without your input.

The sarcomere is the fundamental contractile unit of muscle tissue, and it may range in length from 1.5 μm (full shortening) to 2.5 μm (at rest) and up to 4 μm (full extending). Skeletal muscles, the most prevalent tissue making up 40–50% of the mass of the human body, are mostly responsible for the movement of the skeleton by carrying loads during body motion and are connected to bones [2–6]. Water, fat, and proteins make up the structure of skeletal muscle, including myofibrillar proteins, regulatory proteins (m-protein, gamma-actin, beta-actin, tropomyosin, and troponin), structural proteins (a-actinin, b-actinin, c-protein, h-protein, filamin, myosin, and tropomodulin), and stromal proteins.

The muscular fascicle, muscle fiber, and myofibrils are the structures that make up skeletal muscle. Actin (I-band) and myosin (A-band) filaments are arranged in skeletal muscle in a regular pattern, giving it a striated shape. At the Z-line, myofibrils are connected to one another inside the I-band. Actin, tropomyosin, and troponin proteins make up actin filaments. Myosin proteins make up myosin filaments. Titin is

A. Chanda and G. Singh, *Mechanical Properties of Human Tissues*,
Materials Horizons: From Nature to Nanomaterials,
https://doi.org/10.1007/978-981-99-2225-3_3

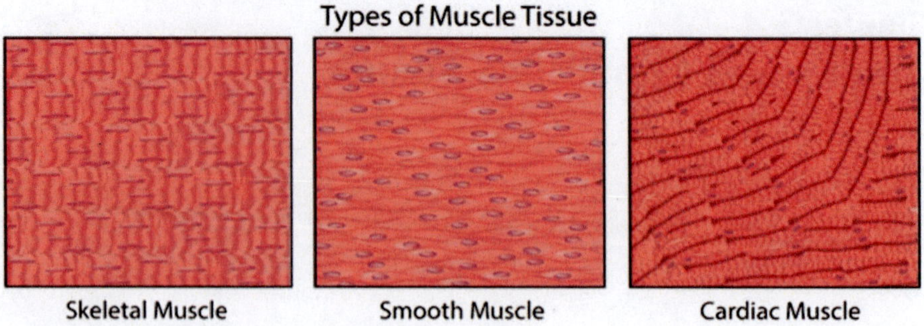

Fig. 3.1 Different types of muscles and their structures [7]

a structural protein that connects the opposing ends of the actin and myosin filaments with an elastic band (Fig. 3.2). The respiratory system, the urogenital tract, and organs including the stomach and intestines are all generally surrounded by smooth muscles [8–11]. Cardiac muscles contract at a certain pace that is inherently regulated by brain impulses and hormones, making them involuntary and self-stimulating.

Muscle fibers (diameter 100 μm), myofibrils (1 μm), and myofilaments (100 Å) make up the three components of muscle tissue, which are together known as fascia. To assess the change in muscular length, or the tension that has generated in the muscle and the isolated muscle is usually stretched to estimate its stiffness [12]. Edwards et al. [13] estimated the average rupture force for the cardiac muscle and reported to be 4.9 N $\pm$ 3.6 N. Sarcomere lengths less than 1.85 mm or more than 2.2 mm significantly affect the overall tension of the muscle tissue [14]. Most studies on the uniaxial mechanical properties of muscle tissues used quasi-static loading and compression testing conditions [15, 16]. Zwirner et al. [17] examined the load-deformation characteristics of the temporal muscle, which is often employed as a graft material in craniofacial surgery [18, 19]. The temporal muscle's mean elastic modulus and ultimate tensile stress were reported to be 1.58 $\pm$ 0.64 MPa and 0.26 $\pm$ 0.11 MPa, respectively. The skeletal muscle was subjected to in vivo magnetic resonance elastography by Uffmann et al. [20]. The biceps brachii, flexor digitorum profundus, soleus, and gastrocnemius shear moduli were determined to be 17.9 $\pm$ 5.5 kPa, 8.7 $\pm$ 2.8 kPa, 12.5 $\pm$ 7.3 kPa, and 9.9 $\pm$ 6.8 kPa, respectively. The Poisson's ratio for the contracted muscles and relaxed muscles was estimated by Payne et al. [1], and the reported values were 0.480 and 0.493, respectively. There is limited literature on the experimental data of the realistic mechanical properties of human muscle tissue. The animal models were extensively used by researchers to estimate the force–velocity relationship, muscle-length tension under controlled experimental settings [21–26]. Table 3.1 shows the comparison of all the muscle types, i.e., skeletal, smooth, and cardiac muscles.

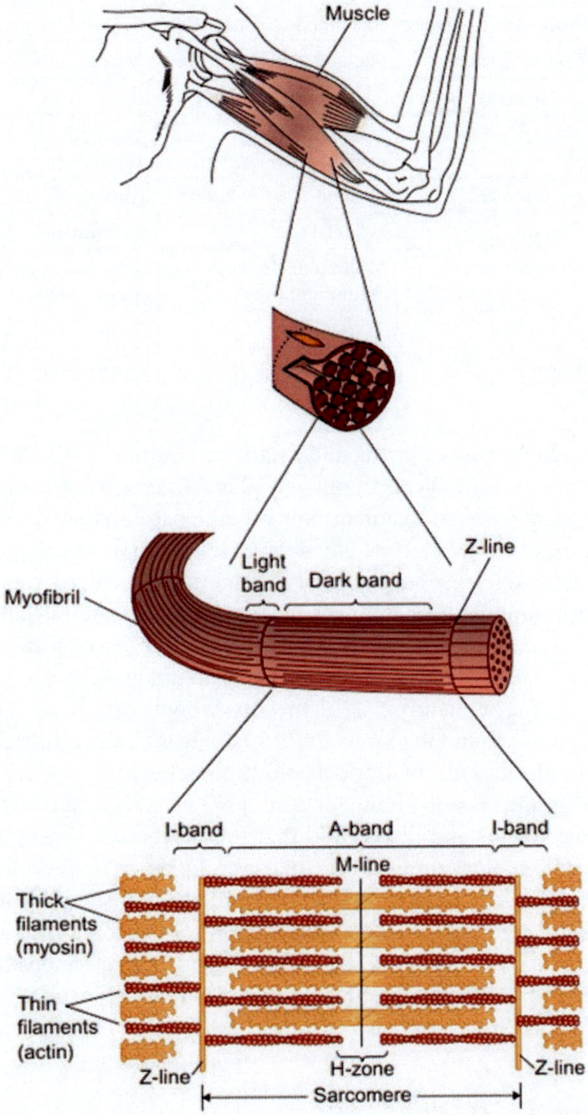

Fig. 3.2 Structure of skeletal muscle [11]

3.2 Connective Tissues

In the connective tissues, elastin is found between and along the collagen fascicles and fibers. Elastin may withstand a significant amount of stress when connective tissues are stretched along the primary collagen axis [27]. Up to 70% of the dry weight of a ligament or tendon is made up of water, and the remaining components include type I collagen, which makes up the majority of the remaining constituents, along with

Table 3.1 Characteristic comparison of different muscle tissue

Characteristic	Skeletal muscle	Smooth muscle	Cardiac muscle
Contraction speed	Fast	Slow	Intermediate
Fiber protein	Actin, myosin, tropomyosin, troponin	Actin, myosin, tropomyosin	Actin, myosin, tropomyosin, troponin
Appearance	Striated	Smooth	Striated
Control	Voluntary	Involuntary	Involuntary
Morphology	Multinucleate, large cylindrical fibers	Uninucleate, small spindle-shaped fibers	Uninucleate, shorter branching fibers
Activation	Troponin	Calmodulin MLCK	Troponin
Calcium source	Intracellular	Extracellular and SR	Extracellular and SR

proteoglycans, elastin, and smaller quantities of fibrillin, fibrinogen, fibronectin, and laminin [28]. Type I collagen plays a significant role in the hierarchical structural organization of ligaments and other dense connective tissues. This hierarchical structure spans a variety of physical scales, from tropocollagen molecules to fibrils, fibers, and fascicles, and ultimately to the macro-scale tissue, where collagen's mesoscale crimping and twisting result in nonlinear material behavior [29]. The other extracellular components of the matrix, or the "ground material," have not been studied extensively, with regard to their structure and function.

Approximately 5% of the dry weight of ligaments is made up of elastin. It is composed in the extracellular space from a fibrillin-rich microfibril scaffold around an elastin core of tropoelastin molecules [30, 31]. By using elastase to selectively degrade elastin, Henniger et al. [27] investigated the function of elastin in ligament mechanics and discovered that it contributed disproportionately to uniaxial tensile deformation along the primary collagen axis. Despite making up just 4% of the tissue's dry weight, elastin was able to withstand up to 30% of tensile stress when subjected to uniaxial strain. Elastin is also found in the cruciate ligaments between and along collagen fibers. Such reasons make the prominence that, in relation to the major collagen axis, elastin may also resist transverse and shear tissue deformation.

3.2.1 Tendons

Tendons are skeletal muscle-to-bone connections that help support or move joints by transmitting the stresses generated by muscle contractions. They begin in the muscle, pass through at least one joint (and sometimes more), and then implant into the bone. Tendons have a well-known function in proprioception. Tendons also provide the vital role of storing energy. Walking enables for effective locomotion due to the phasic transfer of energy between potential and kinetic forms during the up-and-down movements of each stride. Instead, energy efficiency is accomplished by temporarily storing energy as elastic strain energy in stretched tendons and releasing it back into the system later in the gait cycle. This process malfunctions while running, and energy efficiency is instead obtained by doing this. By transmitting the forces generated by muscle contractions to sustain or create motion via joints, tendon offers the coordinating link between a muscle and a bone [32]. The parallel arrangement of the tendon fibers in the direction of the applied force exhibits the uniaxial stress–strain relationship [33]. The tendons transfer muscular forces and possess a high tensile strength of 100–140 MPa with an elastic modulus of 1.0–1.5 GPa [34]. The Achilles tendon, a fibrous structure that connects the calf muscles with the calcaneus, was estimated to have a Poisson's ratio of 0.3 and an elastic modulus of 0.8 GPa [35]. Tendon collagen was reported with an elastic modulus in the range of 3–9 GPa [36]. According to Zitnay and Weiss [37], freezing considerably decreased the elastic modulus of human tendons, with the mean value falling from 244 to 180 MPa.

3.2.2 Ligaments

Ligaments, which hold bones together and support organs, are strong bands of fibrous connective tissue. Their major purpose from a skeletal perspective is to maintain proper bone and joint shape. Ligaments are referred to be passive joint stabilizers because they control a joint's range of motion together with the articular outlines and their accompanying joint capsules, which are made of a similar substance. When the joint is pressed outside of its functional range, ligamentous injury occurs. Ligaments, like tendons, are connective tissues that support the musculoskeletal system and transport stresses across the joints of the skeleton to carry out a variety of physiological activities. Ligaments exhibit mechanical viscoelastic behavior and are made of thick connective fibrous tissue [38]. Ligaments join one bone to another to restrict relative motions. For the mechanical properties of the ligaments, knee ligaments were reported to have an elastic modulus and a Poisson's ratio in the range of 3.75–4.3 GPa and 0.3, respectively [39, 40]. The elastic modulus and Poisson's ratio for the patellar ligament were reported to be 0.4 and 225 MPa, respectively [41], and 0.4 and 260 MPa, respectively, for the ankle ligament [42]. Figure 3.3 shows the elastin network in ligament tissue under different phases, i.e., unloaded, loaded, and elastase.

Together, these connective tissues function as a muscle–bone bridge that executes the appropriate action in response to the signal sent by the brain. Elastin, which makes up around 1–4% of the dry weight of tendons and 4–9% of the dry weight of ligaments, is the main non-collagenous protein found in tendons and ligaments. Similar to elastin, collagen makes about 65–80% of the dry weight and forms a fibrillar network with a hierarchically organized that is predominantly oriented in the loading direction [43, 44]. Tendons and ligaments may be stretched 5–7% without any damage, while a ligament or tendon can typically withstand 12–15% of strain before failing. The tendon experiences the least amount of energy loss and distortion for stretches that are less than 10% of its initial length. Over the intervertebral disk space, the longitudinal ligaments ranged in length from 4.2 to 6.5 mm [45]. According

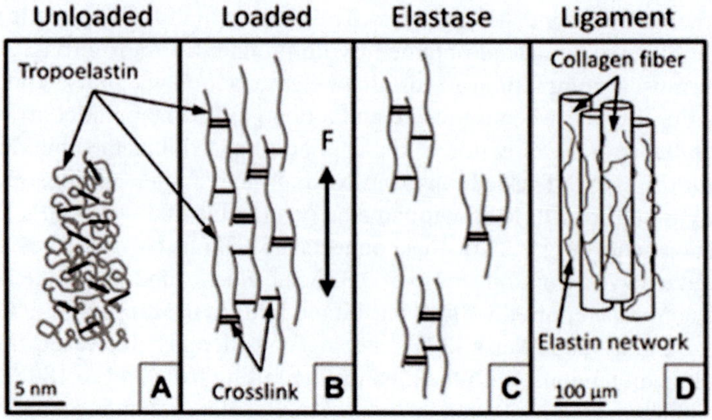

Fig. 3.3 Representation of the elastin network in ligament tissue. **a** In an unloaded state, the crosslinked elastin network exists as an unorganized network of randomly coiled elastin fibers. **b** Under applied load, the elastin network elongates, and **c** following selective digestion with elastase the network is disrupted and only fragments remain. **d** Representation of elastin residing along and between collagen fibers in ligament [36]

to Martin et al. [32], tendons and ligaments have an elastic modulus between 1.0 and 2.0 GPa and an ultimate tensile strength between 50 and 150 MPa.

References

1. Payne T, Mitchell S, Bibb R, Waters M (2015) Development of novel synthetic muscle tissues for sports impact surrogates. J Mech Behav Biomed Mater 41:357–374. https://doi.org/10.1016/j.jmbbm.2014.08.011
2. Mukund K, Subramaniam S (2020) Skeletal muscle: a review of molecular structure and function, in health and disease. Wiley Interdisc Rev Syst Biol Med 12(1). (Wiley-Blackwell). https://doi.org/10.1002/wsbm.1462
3. Takaza M, Moerman KM, Gindre J, Lyons G, Simms CK (2013) The anisotropic mechanical behaviour of passive skeletal muscle tissue subjected to large tensile strain. J Mech Behav Biomed Mater 17:209–220. https://doi.org/10.1016/j.jmbbm.2012.09.001
4. Chomentowski P, Coen PM, Radiková Z, Goodpaster BH, Toledo FGS (2011) Skeletal muscle mitochondria in insulin resistance: differences in intermyofibrillar versus subsarcolemmal subpopulations and relationship to metabolic flexibility. J Clin Endocrinol Metab 96(2):494–503. https://doi.org/10.1210/jc.2010-0822
5. Hashemi SS, Asgari M, Rasoulian A (2020) An experimental study of nonlinear rate-dependent behaviour of skeletal muscle to obtain passive mechanical properties. Proc Inst Mech Eng Part H J Eng Med 234(6):590–602. https://doi.org/10.1177/0954411920909705
6. Lindskog C et al (2015) The human cardiac and skeletal muscle proteomes defined by transcriptomics and antibody-based profiling. BMC Genomics 16(1):475. https://doi.org/10.1186/s12864-015-1686-y
7. Smooth, Skeletal, and Cardiac Muscles. https://bio.libretexts.org/Bookshelves/Introductory_and_General_Biology/Book%3A_Introductory_Biology_%28CK-12%29/13%3A_Human_Biology/13.15%3A_Smooth_Skeletal_and_Cardiac_Muscles. Accessed 27 December 2022

8. Vignos PJ, Lefkowitz M (1959) A biochemical study of certain skeletal muscle constituents in human progressive muscular dystrophy. J Clin Invest 38(6):873–881. https://doi.org/10.1172/JCI103869
9. Van Loocke M, Simms CK, Lyons CG (2009) Viscoelastic properties of passive skeletal muscle in compression-Cyclic behaviour. J Biomech 42(8):1038–1048. https://doi.org/10.1016/j.jbiomech.2009.02.022
10. Boland M, Kaur L, Chian FM, Astruc T (2019) Muscle proteins. Encycl Food Chem 164–179. https://doi.org/10.1016/B978-0-08-100596-5.21602-8
11. Hill JA, Olson EN (2012) An introduction to muscle. In: Muscle, vol 1. Elsevier Inc., pp 3–9
12. Herbert R (1988) The passive mechanical properties of muscle and their adaptations to altered patterns of use. Aust J Physiother 34(3):141–149. https://doi.org/10.1016/S0004-9514(14)60606-1
13. Edwards MB, Draper ERC, Hand JW, Taylor KM, Young IR (2005) Mechanical testing of human cardiac tissue: some implications for MRI safety. J Cardiovasc Magn Reson 7(5):835–840. https://doi.org/10.1080/10976640500288149
14. Hunter PJ, McCulloch AD, Ter Keurs HEDJ (1998) Modelling the mechanical properties of cardiac muscle. Prog Biophys Mol Biol 69(2–3):289–331. https://doi.org/10.1016/S0079-6107(98)00013-3
15. Fung Y-C, Fung Y-C (1993) Introduction: a sketch of the history and scope of the field. In: Biomechanics. Springer, New York, pp 1–22
16. Gras LL, Mitton D, Crevier-Denoix N, Laporte S (2012) The non-linear response of a muscle in transverse compression: assessment of geometry influence using a finite element model. Comput Methods Biomech Biomed Engin 15(1):13–21. https://doi.org/10.1080/10255842.2011.564162
17. Zwirner J, Ondruschka B, Scholze M, Hammer N (2020) Passive load-deformation properties of human temporal muscle. J Biomech 106:109829. https://doi.org/10.1016/j.jbiomech.2020.109829
18. Clauser L, Curioni C, Spanio S (1995) The use of the temporalis muscle flap in facial and cranio-facial reconstructive surgery. a review of 182 cases. J. Cranio-Maxillofac Surg. 23(4):203–214. https://doi.org/10.1016/S1010-5182(05)80209-4
19. Lam D, Carlson ER (2014) The temporalis muscle flap and temporoparietal fascial flap. Oral Maxillofac Surg Clin N Am 26(3):359–369. https://doi.org/10.1016/j.coms.2014.05.004
20. Uffmann K et al (2004) In vivo elasticity measurements of extremity skeletal muscle with MR elastography. NMR Biomed 17(4):181–190. https://doi.org/10.1002/nbm.887
21. Bosboom EMH, Hesselink MKC, Oomens CWJ, Bouten CVC, Drost MR, Baaijens FPT (2001) Passive transverse mechanical properties of skeletal muscle under in vivo compression. J Biomech 34(10):1365–1368. https://doi.org/10.1016/S0021-9290(01)00083-5
22. Hawkins D, Bey M (1997) Muscle and tendon force-length properties and their interactions in vitro. J Biomech 30(1):63–70. https://doi.org/10.1016/S0021-9290(96)00094-2
23. Gareis H, Moshe S, Baratta R, Best R, D'Ambrosia R (1992) The isometric length-force models of nine different skeletal muscles. J Biomech 25(8):903–916. https://doi.org/10.1016/0021-9290(92)90230-X
24. Woittiez RD, Huijing PA, Boom HBK, Rozendal RH (1984) A three-dimensional muscle model: a quantified relation between form and function of skeletal muscles. J Morphol 182(1):95–113. https://doi.org/10.1002/jmor.1051820107
25. Muhl ZF (1982) Active length-tension relation and the effect of muscle pinnation on fiber lengthening. J Morphol 173(3):285–292. https://doi.org/10.1002/jmor.1051730305
26. Gordon AR, Siegman MJ (1971) Mechanical properties of smooth muscle. I. Length-tension and force-velocity relations. Am J Physiol 221(5):1243–1249. https://doi.org/10.1152/ajplegacy.1971.221.5.1243
27. Henninger HB, Valdez WR, Scott SA, Weiss JA (2015) Elastin governs the mechanical response of medial collateral ligament under shear and transverse tensile loading. Acta Biomater 25:304–312. https://doi.org/10.1016/j.actbio.2015.07.011

28. Halper J, Kjaer M (2014) Basic components of connective tissues and extracellular matrix: elastin, fibrillin, fibulins, fibrinogen, fibronectin, laminin, tenascins and thrombospondins, pp 31–47. https://doi.org/10.1007/978-94-007-7893-1_3/COVER
29. Reese SP, Maas SA, Weiss JA (2010) Micromechanical models of helical superstructures in ligament and tendon fibers predict large Poisson's ratios. J Biomech 43(7):1394–1400. https://doi.org/10.1016/J.JBIOMECH.2010.01.004
30. Kielty CM, Sherratt MJ, Shuttleworth CA (2002) Elastic fibres. J Cell Sci 115(14):2817–2828. https://doi.org/10.1242/JCS.115.14.2817
31. Henninger HB, Underwood CJ, Romney SJ, Davis GL, Weiss JA (2013) Effect of elastin digestion on the quasi-static tensile response of medial collateral ligament. J Orthop Res 31(8):1226–1233. https://doi.org/10.1002/JOR.22352
32. Martin RB, Burr DB, Sharkey NA, Martin RB, Burr DB, Sharkey NA (1998) Mechanical properties of ligament and tendon. In: Skeletal tissue mechanics. Springer, New York, pp 309–348
33. Carniel TA, Formenton ABK, Klahr B, Vassoler JM, de Mello Roesler CR, Fancello EA (2019) An experimental and numerical study on the transverse deformations in tensile test of tendons. J Biomech 87:120–126. https://doi.org/10.1016/j.jbiomech.2019.02.028
34. Biewener AA (2008) Tendons and ligaments: structure, mechanical behavior and biological function. In Collagen: structure and mechanics. Springer US, pp 269–284
35. Wren TAL, Yerby SA, Beaupré GS, Carter DR (2001) Mechanical properties of the human achilles tendon. Clin Biomech 16(3):245–251. https://doi.org/10.1016/S0268-0033(00)00089-9
36. Zitnay JL, Weiss JA (2018) Load transfer, damage, and failure in ligaments and tendons. J Orthop Res 36(12):3093–3104. https://doi.org/10.1002/jor.24134
37. Clavert P, Kempf JF, Bonnomet F, Boutemy P, Marcelin L, Kahn JL (2001) Effects of freezing/thawing on the biomechanical properties of human tendons. Surg Radiol Anat 23(4):259–262. https://doi.org/10.1007/s00276-001-0259-8
38. Woo SLY, Orlando CA, Camp JF, Akeson WH (1986) Effects of postmortem storage by freezing on ligament tensile behavior. J Biomech 19(5):399–404. https://doi.org/10.1016/0021-9290(86)90016-3
39. Mo F, Li F, Behr M, Xiao Z, Zhang G, Du X (2018) A Lower limb-pelvis finite element model with 3D active muscles. Ann Biomed Eng 46(1):86–96. https://doi.org/10.1007/s10439-017-1942-1
40. Mo F, Arnoux PJ, Cesari D, Masson C (2012) The failure modelling of knee ligaments in the finite element model. Int J Crashworthiness 17(6):630–636. https://doi.org/10.1080/13588265.2012.704194
41. Butler DL, Kay MD, Stouffer DC (1986) Comparison of material properties in fascicle-bone units from human patellar tendon and knee ligaments. J Biomech 19(6):425–432. https://doi.org/10.1016/0021-9290(86)90019-9
42. Siegler S, Schneck CD (1988) The mechanical characteristics of the collateral ligaments of the human ankle joint. Foot Ankle Int 8(5):234–242. https://doi.org/10.1177/107110078800800502
43. Buckley MR et al (2013) Distributions of types I, II and III collagen by region in the human supraspinatus tendon. Connect Tissue Res 54(6):374–379. https://doi.org/10.3109/03008207.2013.847096
44. Wu F, Nerlich M, Docheva D (2017) Tendon injuries: basic science and new repair proposals. EFORT Open Rev. 2(7):332–342. https://doi.org/10.1302/2058-5241.2.160075
45. Yoganandan N, Kumaresan S, Pintar FA (2000) Geometric and mechanical properties of human cervical spine ligaments. J Biomech Eng 122(6):623–629. https://doi.org/10.1115/1.1322034

Chapter 4
Tissues in Functional Organs—Low Stiffness

4.1 Brain

The human brain tissue is a soft wrinkled organ located within the cranial cavity and protected by the bones of the skull, and its activity has been confirmed to be nonlinear and viscoelastic in nature [1, 2]. The brain, which weighs roughly three pounds and has around 100 billion neurons, is the central nervous system of the human body. It is composed of the intracranial cerebrum and the cerebellum, and the spinal cord, which transmits information to the rest of the body. The meninges are layers of membranes that surround and protect the brain, allowing it to be filled with and submerged in cerebrospinal fluid. The cerebral cortex may be distinguished physically from the rest of the brain by the presence of white matter atop the outermost gray matter [3, 4]. Among the gray matter's constituents are data-processing neurons, while the white matter's constituents include myelinated nerve fiber bundles that promote quick signal transmission [1, 5].

Because of the intricacy of the experimental testing, such as holding the sample, compression studies are often used to examine the mechanical characteristics of brain tissues [6]. Tensile tests are rarely used to characterize the mechanical properties of brain tissues. The tensile characteristics of brain tissue have only been described in a few studies. Despite the fact that failure limits emerge in the region of 20–60% strain [7], they are not well understood. Figure 4.1 shows the biomechanical testing of human brain tissue during simple shear, compressive, and tensile loadings. Zhang et al. [8] reported a significant reduction in stiffness in both the white matter and gray matter of the brain as a result of the prolonged storage duration. The engineering stresses in the white matter decrease from 1.68 $\pm$ 0.31 kPa to 0.74 $\pm$ 0.22 kPa, whereas the engineering stresses in the gray matter decrease from 1.06 $\pm$ 0.41 kPa to 0.69 $\pm$ 0.14 kPa. The white matter of the brain tissue was estimated to have an average shear modulus of 1.01 kPa, which was found to be stiffer than the gray matter, which had a shear modulus of 0.752 kPa (Fig. 4.2) [9]. Huang et al. [10] evaluated the elastic modulus of white matter and gray matter in the brain tissue at frequencies

A. Chanda and G. Singh, *Mechanical Properties of Human Tissues*,
Materials Horizons: From Nature to Nanomaterials,
https://doi.org/10.1007/978-981-99-2225-3_4

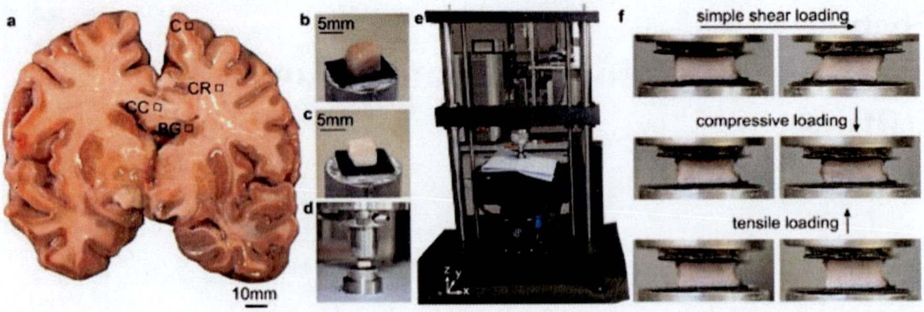

Fig. 4.1 Biomechanical testing of human brain tissue **a** specimens taken from four different brain regions, the corpus callosum (CC), the corona radiata (CR), the basal ganglia (BG), and the cortex (C); **b** gray matter specimen; **c** white matter specimen; **d** white matter specimen mounted to the testing device; **e** triaxial biomechanical testing device; **f** close ups of the tissue deformation during simple shear, compressive, and tensile loadings [2]

of 40, 50, and 60 Hz and found that they were significantly different. The respective modulus values for white matter were 3.36 ± 0.111 kPa, 3.78 ± 0.15 kPa, and 3.85 ± 0.12 kPa, whereas the corresponding modulus values for gray matter were 2.24 ± 0.14 kPa, 2.82 ± 0.16 kPa, and 3.33 ± 0.14 kPa, respectively. On the other hand, Budday et al. [11] reported the average value for the elastic modulus of brain tissue. The estimated elastic modulus of white matter was found to be 1.895 ± 0.592 kPa, which was higher than the 1.389 ± 0.289 kPa found for gray matter. In a study by MacManus et al. [12], they found that the average cortical thickness of the human was 2622 μm from the pial layer to white matter. It is reported in the literature that the human brain tissue has no characteristic differences from the porcine, bovine, murine, or ovine brain tissue [13].

Another study experimentally tested the brain tissue under compression loading conditions and the compressive modulus of human brain tissue was found to be 10.34 kPa [14]. Various strain rates of 30, 60, and 90/s were used to determine the nominal compressive stress, which was found to be 8.83 ± 1.944 kPa, 12.8 ± 3.10 kPa, and 16.0 ± 1.41 kPa, respectively [5]. Chatelin et al. [15] estimated the incompressibility of brain tissue and the corresponding complex shear modulus value of 24.2 kPa. Compared to gray matter, white matter in the brain tissue exhibited higher anisotropy and, when subjected to higher strain rates, exhibited 1.3 times more stiffness [15, 16]. To compare human brain tissue with porcine brain tissue, Prange and Margulies [16] used the shear stress relaxation approach. They found that human brain tissue was 29% stiffer than pig brain tissue. Similar results were seen in human brain samples, which were 40% stiffer than bovine brain samples [17]. According to the findings of a research study conducted on human subjects using the magnetic resonance elastography method (age range 12–72 years) [18], the shear modulus of brain tissue decreases as the subject's age increases. Yeung et al. [19] conducted a similar kind of investigation on human volunteers, finding that the stiffness of the brain tissue (both white and gray matter) was almost identical in both children (7–12 years) and adults (> 18 years).

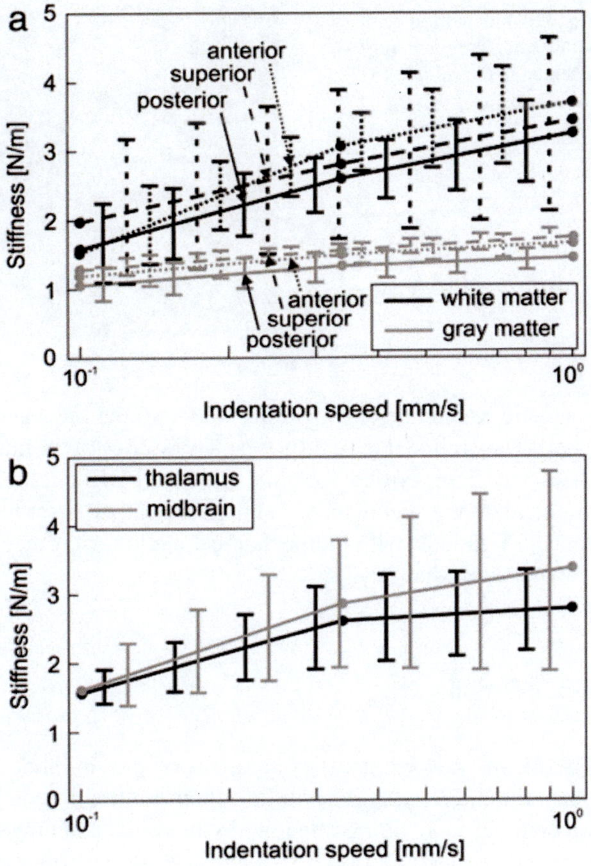

Fig. 4.2 Indentation stiffness at 0.1 mm versus indentation speed for **a** white and gray matter tissues of the posterior, superior, and anterior regions of the cerebrum and **b** thalamus and midbrain tissue [9]

4.2 Tongue

The tongue is a functional organ of the upper airway, aiding in swallowing, speaking, licking, and breathing. It is composed of a rough, thin, skin-like coating of many muscles [20]. The tongue is located posterior and medial to the teeth and functions to grab and force food into the mouth through the tongue's muscle [21, 22]. On the tongue's surface, the taste buds detect food flavor, and the information is then conveyed to the brain.

Although some works in the literature cover biomedical modeling of the human tongue [20, 21, 23], the mechanical characteristics of the tongue need considerable investigation. A few studies have calculated the modulus of elasticity for the base of human tongue tissue to be between 6 kPa [24–26] and 15 kPa [22]. The elastic modulus of the human tongue was recently determined to be 5.52 ± 1.19 kPa using the indentation method [27]. Figure 4.3 shows the indentation device with human tongue being indented in the middle. Brown et al. [28] and Cheng et al. [29] used

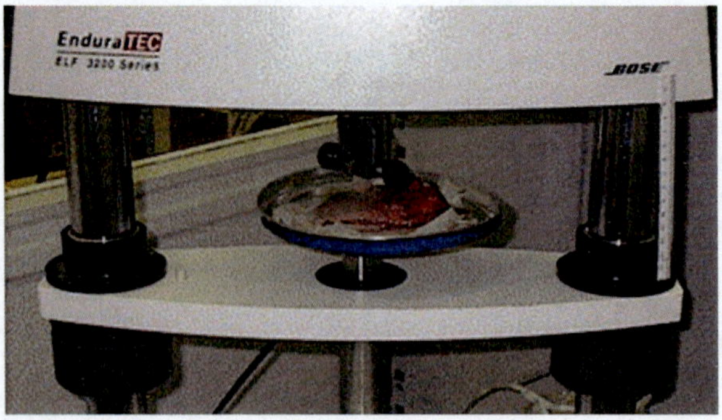

Fig. 4.3 Indentation device with human tongue being indented in the middle [31]

magnetic resonance elastography to determine the shear modulus of human tongue tissue. They found it to be 2.68 kPa. Knowledge of the human tongue's biomechanical characteristics is critical for finite element (FE) modeling of the upper airway in order to elucidate the mechanism behind respiratory disorders such as obstructive sleep apnea. FE models with biomechanical features may also forecast speech therapy and jaw repair results [30–32].

4.3 Tonsils

The tonsils are a critical component of our immune system, limiting pathogens (bacteria and viruses) and killing them with antibodies made by immune cells as we breathe, so avoiding infection in the throat and lungs. Each side of the mouth has two tonsils (fleshy lumps). Tonsils vary in size from person to person, and they are often inflamed after infection. Obstructive sleep apnea (OSA) is a chronic condition characterized by abnormal upper airway deformation, enlarged tonsils, and other oropharyngeal tissues such as the tongue, soft palate, and uvula.

The mechanical properties of the tonsils are still limited. Figure 4.4 shows the indentation of freshly excised human tonsil tissue. This soft tissue needs investigation because they are important for modeling accurate and reliable biomechanical models for upper airway collapse simulations used to analyze OSA situations [27]. Additionally, such FEM studies, considering the realistic mechanical properties of the tissue, promote minimal invasive surgical treatments. However, the literature has rarely reported the shear modulus of elasticity of human tonsils. Haddad et al. [27] recently evaluated the elastic modulus of fresh human tonsils and determined it to be 4.56 ± 2.41 kPa.

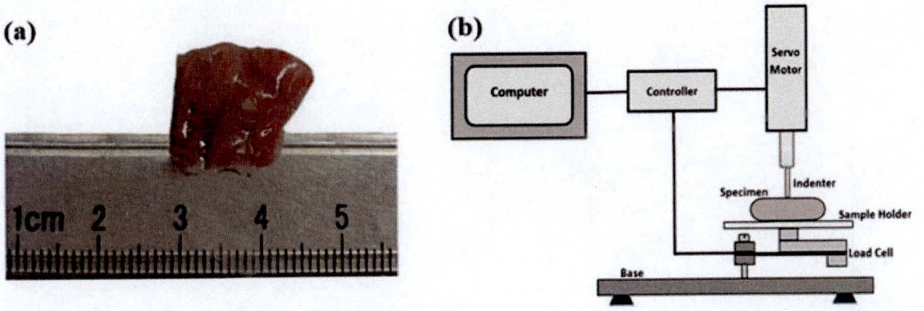

Fig. 4.4 **a** Tonsil specimen for the indentation test and **b** schematic of the indentation system for acquiring tissue indentation force–displacement data [27]

4.4 Esophagus

The esophagus is a nonlinear muscular tube about 25–30 cm long and 2–3 cm thick that delivers food and water to the stomach through the neck region (pharynx). The cardiac sphincter, a muscular ring at the esophagus's inferior end, seals the esophagus and captures food in the stomach [33, 34]. The esophagus wall has active and passive biomechanical properties, i.e., muscle contractions (active) and elastic and viscoelastic characteristics (passive) [35]. Egorov et al. [36] used a tensiometer to determine the tensile characteristics of the human esophagus (Fig. 4.5). They found ultimate stress of 1.2 MPa and a destructive strain of 140%. There is limited literature published on the mechanical characteristics of human esophageal tissue, such as elastic modulus, shear modulus, and ultimate tensile strength. Orvar et al. [37] determined the pressure elastic modulus of the human esophagus, which varies between 4 and 14 kPa. Yang et al. [38] claimed to have developed the first method for determining the esophageal shear modulus using an animal model (rat). The calculated shear modulus ranges between 5.43 and 185.01 kPa at a longitudinal length of 1.5 and between 0 and 2 kPa inflation pressures.

4.5 Lungs

The lung tissue is a pair of big, spongy organs placed lateral to the heart and superior to the diaphragm in the thorax. Due to the direction of the heart, the left lung is somewhat smaller than the right lung and has two lobes, while the right lung has three lobes. The lung's primary work is to provide oxygen to the blood and to cleanse it of carbon dioxide in exchange (via tiny air sacs called alveoli). Jawde et al. [39] recently found that the elastic modulus of the alveolar wall is age-dependent and increases with age from 4.4 to 5.9 kPa. The elastic moduli of fibers inside the alveolar wall followed a similar pattern, ranging from 311 to 620 kPa. The mechanical properties of lung tissue are highly dependent on the extracellular matrix fibrous proteins,

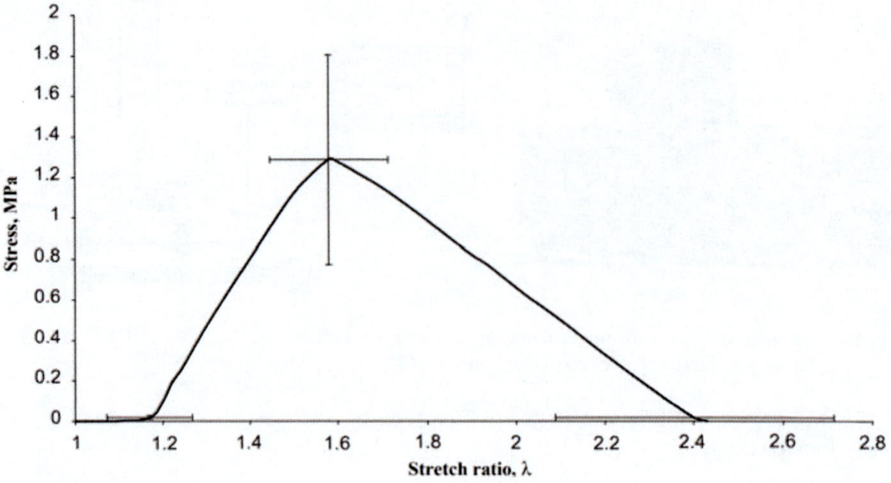

Fig. 4.5 Mechanical testing of human cadaveric esophagus axial specimens [36]

namely collagen and elastin [40]. Within a 20 μm alveolar duct, the densities of elastin and collagen fibers in human lung samples are 16% and 18%, respectively [41]. Another study determined the elastin and collagen concentration to be 15% and 29%, respectively [42]. Polio et al. [64] estimated the lung tissue's elastic modulus using cavitation rheology (6.1 ± 1.6 kPa), micro-indentation (1.4 ± 0.4 kPa), and uniaxial (3.4 ± 0.4 kPa) tests. Booth et al. [43] examined human cadaver lung tissue and determined the elastic modulus of native normal lung and normal lung that had been decellularized. In both instances, elastic moduli were reported to be 1.96 ± 0.13 kPa and 1.6 ± 0.08 kPa. Similar investigations have been conducted and have determined the modulus of elasticity and Poisson's ratio of human lung tissue to be 5 kPa and 0.22, respectively [44, 45]. Due to the unavailability of the cadaveric lung tissue, limited studies have performed experimental testing of the human lung tissue. In the recent studies, the computational modeling of lung tissue is of great interest where different image tools (CT scan, MRI, etc.) are being used to model the realistic lung model (Fig. 4.6) [44].

4.6 Breast

Fat, fibrous tissue, and functioning glandular tissues such as mammary glands and milk ducts comprise the breast [46, 47]. During breastfeeding, the infant gets the milk from the milk collecting ducts, which open at the nipple. The breast is divided into 15–25 lobes by branching ducts connecting the collecting ducts to the terminal lobular units. The thickness of the breast skin is critical in defining the mechanical properties of the breast tissue. The mean thickness of breast skin described in the literature was

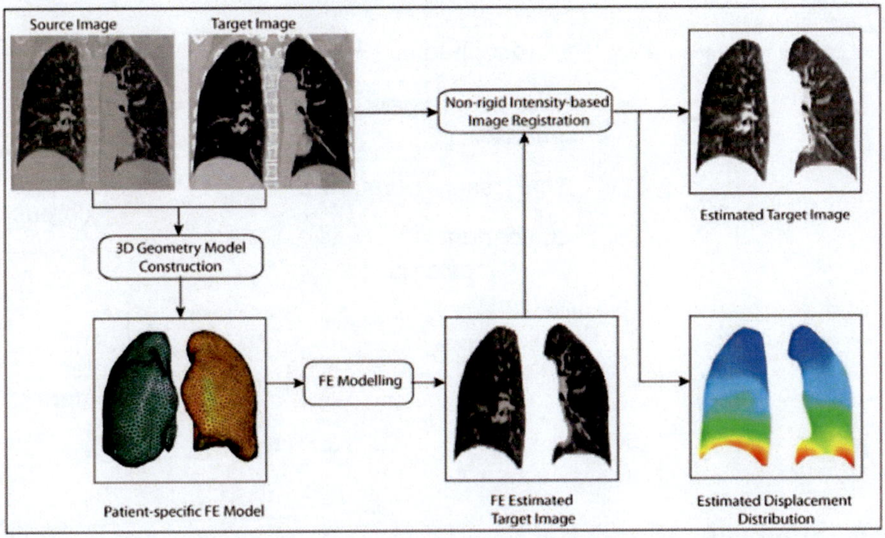

Fig. 4.6 Patient-specific biomechanical modeling of lung tissue [44]

1.55 ± 0.25 mm for ultrasound [48], 1.32 ± 0.33 mm for mammography [49], and 1.45 ± 0.30 mm for computed tomography [50]. Additionally, the breast skin thickness was assessed to be between 0.50 mm and 2.10 mm [48–52].

Understanding the elastic characteristics of breast tissue may aid in a variety of medical applications, including determining the changes due to radiation and tissue expansion during implant-based breast restoration [53]. Sutradhar and Miller [48] evaluated the range for the elastic modulus of breast tissue and reported it to be within the value 195 kPa to 480 kPa. They declared an average elastic modulus of 344 ± 88 kPa for the breast tissue. Multiple studies have shown that the elastic modulus may be used to calculate the mechanical properties of breast tissue [53–55]. According to Gefen and Dilmoney [56], the second to sixth ribs offer strong structural support for the breast, with mean ultimate stress of 106 MPa and a mean elastic modulus of 11.5 GPa in the lateral/medial direction. Breast tissue has a Poisson's ratio of ~0.5 and is stated to be incompressible [46]. The modulus of elasticity of breast glandular tissue is between 2 and 66 kPa, whereas the shear modulus of adipose tissue is between 0.5 and 25 kPa [57]. In comparison, the elastic modulus of human chest skin estimated between 0.2 and 3 MPa, with a failure strain of 60–75% and a thickness of ~2 mm. Arroyo et al. [58] used vibro-elastography to determine the elastic modulus of cancerous tumors, both benign and malignant. The elastic modulus of benign tumors (39.4 ± 12 kPa) was found to be 65% greater than that of healthy breast tissue (23.91 ± 4.57 kPa). In comparison, a 131% rise in malignant lesions was detected with an elastic modulus of 55.4 ± 7.02 kPa.

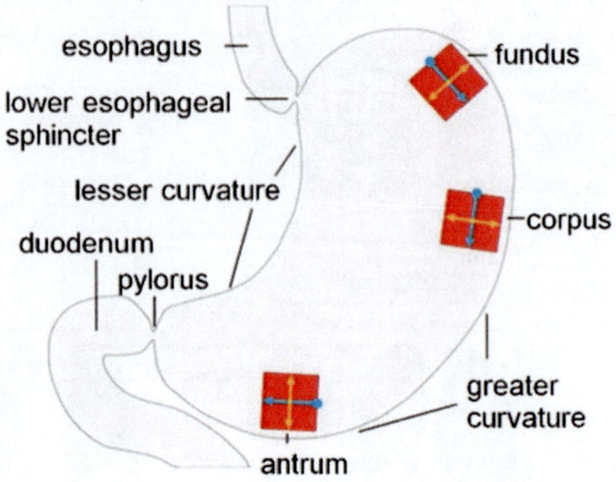

Fig. 4.7 Schematic illustration of the stomach tissue [60]

4.7 Stomach

The stomach is a muscular sac located in the upper abdomen between the esophagus and the duodenum. It is the major organ of the gastrointestinal system and acts as a store for nutrients, ensuring that large meals are absorbed at the correct time. Gastric enzymes and fluids turn partly digested food into a semi-fluid substance in the stomach [59]. Figure 4.7 shows the schematic illustration of the stomach tissue, highlighting the three major regions (i.e., antrum, corpus, and fundus) and duodenum, lower esophageal sphincter, etc. The majority of research on stomach tissue has been conducted on animal models [60–63], and only a few studies have examined the biomechanical properties of human tissue. Egorov et al. [36] studied the axial and transverse tensile characteristics of the human cadaver stomach (Figs. 4.8 and 4.9). At a destructive strain of 190%, they recorded an ultimate stress value of 0.7 MPa and 0.5 MPa, respectively. Lim et al. [64] evaluated the elastic modulus of the human cadaver stomach in situ and reported a mean elastic modulus value of 5.93 kPa for indentation depths ranging from 1.25 ± 0.28 kPa to 3.01 ± 1.30 kPa.

4.8 Spleen

The spleen is a flattened, oval-shaped soft tissue located in the upper left abdomen, lateral to the stomach. It is around 10 cm in length and weighs 150 g. The spleen has a large blood reserve that functions to filter large amounts of blood. It is the least stiff organ in the human abdomen, consisting of a thick fibrous connective tissue capsule filled with red and white pulp [65].

Due to the difficulties of conducting experiments, there is a lack of literature on the mechanical properties of human spleen tissue [66]. Stingl et al. [67] utilized sled

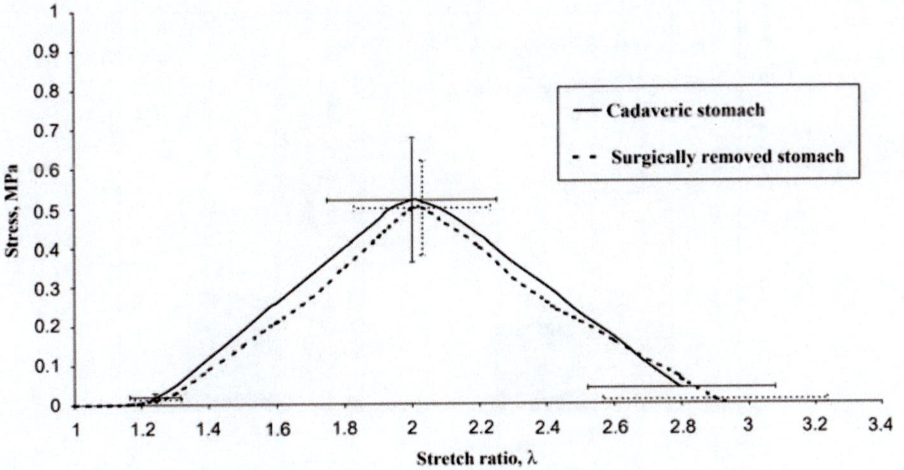

Fig. 4.8 Mechanical testing of human cadaveric and surgically removed transversal stomach specimens [36]

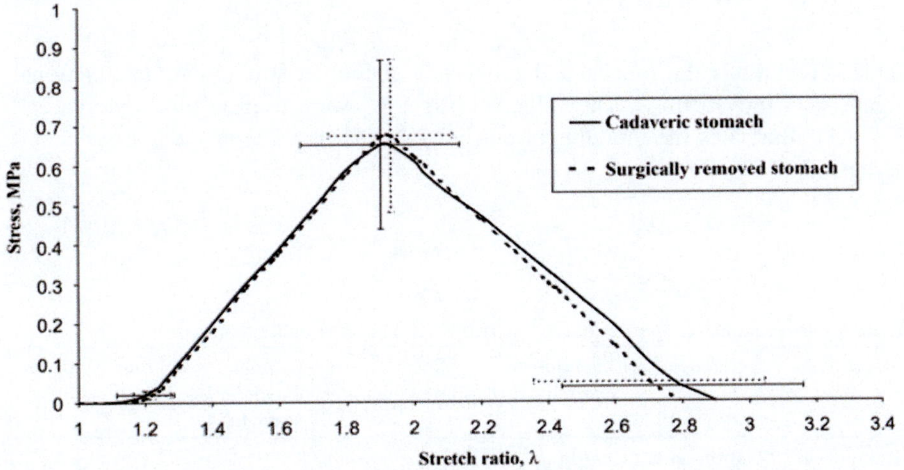

Fig. 4.9 Mechanical testing of human cadaveric and surgically removed axial stomach specimens [36]

tests to determine the mechanical stiffness of fresh cadaver spleen samples. They discovered a range of 0.022–0.652 MPa for male samples and 0.088–0.652 MPa for female samples. Umale et al. [66] estimated the elastic modulus of spleen tissue under static compression to be 14 ± 1.8 kPa for low strain and 35 ± 11 kPa for high strain, respectively. Kemper et al. [65] determined that the spleen tissue's failure stress and strain varied between 26 kPa to 117.3 kPa and 11% to 27%, respectively (Fig. 4.10).

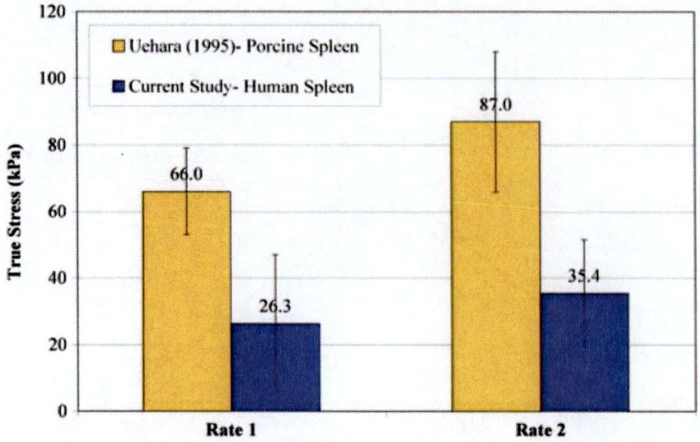

Fig. 4.10 Comparison of tensile failure stress between human and porcine spleen [65]

4.9 Summarizing Mechanical Properties of Tissues with Low Stiffness

Table 4.1 outlines the mechanical properties of human soft tissues in functional organs with low stiffness. The values of Table 4.1 were used to plot a graph, and Fig. 4.11 illustrates the modulus of elasticity (mean ± S.D.) of the discussed soft tissues.

Table 4.1 Mechanical properties of the low stiff tissues in functional organs

Tissue	Reference	Modulus of elasticity	Shear modulus	Ultimate tensile strength	Subgroup/Remarks
Brain tissue	Galford et al. [68]	1–7 kPa	–	–	–
	Green et al. [3]	–	2.7 kPa and 3.1 kPa	–	White matter and gray matter
	Jaw et al. [9]	–	1.01 kPa and 0.752 kPa	–	White matter and gray matter
	Huang et al. [10]	3.66 ± 0.12 kPa and 2.80 ± 0.14 kPa	–	–	White matter and gray matter
	Budday et al. [11]	1.89 ± 0.59 kPa and 1.39 ± 0.29 kPa	–	–	White matter and gray matter

(continued)

Table 4.1 (continued)

Tissue	Reference	Modulus of elasticity	Shear modulus	Ultimate tensile strength	Subgroup/Remarks
Tongue	Malhotra et al., Huang et al., Xu et al. [24–26]	6 kPa	–	–	–
	Payan et al. [22]	15 kPa	–	–	–
	Haddad et al. [27]	5.52 ± 1.19 kPa	–	–	–
	Brown et al. [28]	–	2.68 kPa	–	–
Tonsils	Haddad et al. [27]	4.56 ± 2.41 kPa	–	–	–
Esophagus	Egorov et al. [36]	–	–	1.2 MPa	–
	Orvar et al. [37]	4–14 kPa	–	–	–
Lung tissue	Bou et al. [39]	4.4–5.9 kPa	–	–	Alveolar wall
	Booth et al. [43]	1.96 ± 0.13 kPa	–	–	–
	Han et al. [44]	5 kPa	–	–	–
	Polio et al. [69]	6.1 ± 1.6 kPa	–	–	–
	Polio et al. [69]	1.4 ± 0.4 kPa	–	–	–
	Polio et al. [69]	3.4 ± 0.4 kPa	–	–	–
Breast tissue	Van-Houten et al. [57]	2–66 kPa	–	–	Glandular tissue
	Van-Houten et al. [57]	–	0.5–25 kPa	–	Adipose tissue
	Krouskop et al. [47]	18 ± 7 to 22 ± 12 kPa	–	–	Normal fat*
		28 ± 14 to 35 ± 14 kPa	–	–	Glandular tissue*
		96 ± 34 to 116 ± 28 kPa	–	–	Fibrous tissue*

(continued)

Table 4.1 (continued)

Tissue	Reference	Modulus of elasticity	Shear modulus	Ultimate tensile strength	Subgroup/Remarks
	Krouskop et al. [47]	20 ± 8 to 24 ± 6 kPa	–	–	Normal fat**
		48 ± 15 to 66 ± 17 kPa	–	–	Glandular tissue**
		218 ± 87 to 244 ± 85 kPa	–	–	Fibrous tissue**
Stomach	Egorov et al. [36]	–	–	0.7 MPa	Axial loading
	Egorov et al. [36]	–	–	0.5 MPa	Transversal loading
	Lim et al. [64]	5.93 kPa	–	–	–
Spleen	Stingl et al. [67]	20 kPa	–	–	–
	Umale et al. [66]	24.5 ± 6.4 kPa	–	–	–

Note In Breast tissue: * corresponds 5% precompression and ** corresponds 20% precompressions

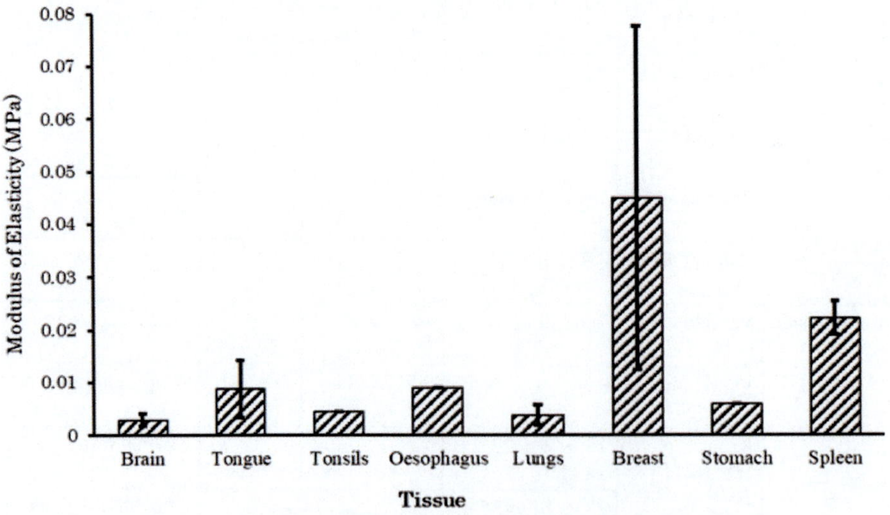

Fig. 4.11 Modulus of elasticity (mean ± S.D.) of the low stiff tissues in functional organs

References

1. Budday S, Sommer G, Birkl C, Langkammer C, Haybaeck J, Kohnert J et al (2017) Mechanical characterization of human brain tissue. Acta Biomater 48:319–340. https://doi.org/10.1016/j.actbio.2016.10.036
2. Budday S, Sarem M, Starck L, Sommer G, Pfefferle J, Phunchago N et al (2020) Towards microstructure-informed material models for human brain tissue. Acta Biomater 104:53–65. https://doi.org/10.1016/j.actbio.2019.12.030
3. Green MA, Bilston LE, Sinkus R (2008) In vivo brain viscoelastic properties measured by magnetic resonance elastography. NMR Biomed 21:755–764. https://doi.org/10.1002/nbm.1254
4. Chanda A, Callaway C, Clifton C, Unnikrishnan V (2018) Biofidelic human brain tissue surrogates. Mech Adv Mater Struct 25:1335–1341. https://doi.org/10.1080/15376494.2016.1143749
5. Rashid B, Destrade M, Gilchrist MD (2012) Mechanical characterization of brain tissue in compression at dynamic strain rates. J Mech Behav Biomed Mater 10:23–38. https://doi.org/10.1016/j.jmbbm.2012.01.022
6. Zhu Z, Jiang C, Jiang H (2019) A visco-hyperelastic model of brain tissue incorporating both tension/compression asymmetry and volume compressibility. Acta Mech 230:2125–2135. https://doi.org/10.1007/s00707-019-02383-1
7. Bilston LE (2011) Brain tissue mechanical properties. Springer, New York, pp 69–89. https://doi.org/10.1007/978-1-4419-9997-9_4
8. Zhang W, Liu L, Xiong Y, Liu Y, Yu S, Wu C et al (2018) Effect of in vitro storage duration on measured mechanical properties of brain tissue. Sci Rep 8. https://doi.org/10.1038/s41598-018-19687-2
9. van Dommelen JAW, van der Sande TPJ, Hrapko M, Peters GWM (2010) Mechanical properties of brain tissue by indentation: Interregional variation. J Mech Behav Biomed Mater 3:158–166. https://doi.org/10.1016/j.jmbbm.2009.09.001
10. Huang X, Chafi H, Matthews KL, Carmichael O, Li T, Miao Q et al (2019) Magnetic resonance elastography of the brain: a study of feasibility and reproducibility using an ergonomic pillow-like passive driver. Magn Reson Imaging 59:68–76. https://doi.org/10.1016/j.mri.2019.03.009
11. Budday S, Nay R, de Rooij R, Steinmann P, Wyrobek T, Ovaert TC et al (2015) Mechanical properties of gray and white matter brain tissue by indentation. J Mech Behav Biomed Mater 46:318–330. https://doi.org/10.1016/j.jmbbm.2015.02.024
12. MacManus DB, Menichetti A, Depreitere B, Famaey N, Vander Sloten J, Gilchrist M (2020) Towards animal surrogates for characterising large strain dynamic mechanical properties of human brain tissue. Brain Multiphysics 1:100018. https://doi.org/10.1016/j.brain.2020.100018
13. Singh G, Chanda A (2021) Mechanical properties of whole-body soft human tissues: a review. Biomed Mater 16:062004. https://doi.org/10.1088/1748-605X/AC2B7A
14. Shuck LZ, Advani SH (1972) Rheologioal response of human brain tissue in shear. J Fluids Eng Trans ASME 94:905–911. https://doi.org/10.1115/1.3425588
15. Chatelin S, Constantinesco A, Willinger R (2010) Fifty years of brain tissue mechanical testing: from in vitro to in vivo investigations. Biorheology 47:255–276. https://doi.org/10.3233/BIR-2010-0576
16. Prange MT, Margulies SS (2002) Regional, directional, and age-dependent properties of the brain undergoing large deformation. J Biomech Eng 124:244–252. https://doi.org/10.1115/1.1449907
17. Takhounts EG, Crandall JR, Darvish K (2003) On the importance of nonlinearity of brain tissue under large deformations. SAE Tech Pap (SAE International). https://doi.org/10.4271/2003-22-0005
18. Sack I, Streitberger K-J, Krefting D, Paul F, Braun J (2011) The influence of physiological aging and atrophy on brain viscoelastic properties in humans. PLoS ONE 6:e23451. https://doi.org/10.1371/journal.pone.0023451

19. Yeung J, Jugé L, Hatt A, Bilston LE (2019) Paediatric brain tissue properties measured with magnetic resonance elastography. Biomech Model Mechanobiol 18:1497–1505. https://doi.org/10.1007/s10237-019-01157-x
20. Hermant N, Perrier P, Payan Y (2017) Human tongue biomechanical modeling. Biomech Living Organs Hyperelastic Const Laws Finite Elem Model 395–411 (Elsevier Inc.). https://doi.org/10.1016/B978-0-12-804009-6.00019-5
21. Kajee Y, Pelteret J-PV, Reddy BD (2013) The biomechanics of the human tongue. Int J Numer Method Biomed Eng 29:492–514. https://doi.org/10.1002/cnm.2531
22. Payan Y, Bettega G, Raphaël B (1998) A biomechanical model of the human tongue and its clinical implications. Lect Notes Comput Sci (Include Subser Lect Notes Artif Intell Lect Notes Bioinform) 1496:688–695 (Springer, Berlin). https://doi.org/10.1007/bfb0056255
23. Gérard JM, Ohayon J, Luboz V, Perrier P, Payan Y (2004) Indentation for estimating the human tongue soft tissues constitutive law: application to a 3D biomechanical model. Lect Notes Comput Sci (Include Subser Lect Notes Artif Intell Lect Notes Bioinform) 3078:77–83. https://doi.org/10.1007/978-3-540-25968-8_9
24. Malhotra A, Huang Y, Fogel RB, Pillar G, Edwards JK, Kikinis R et al (2002) The male predisposition to pharyngeal collapse: importance of airway length. Am J Respir Crit Care Med 166:1388–1395. https://doi.org/10.1164/rccm.2112072
25. Huang Y, White DP, Malhotra A (2007) Use of computational modeling to predict responses to upper airway surgery in obstructive sleep apnea. Laryngoscope 117:648–653. https://doi.org/10.1097/MLG.0b013e318030ca55
26. Xu C, Brennick MJ, Dougherty L, Wootton DM (2009) Modeling upper airway collapse by a finite element model with regional tissue properties. Med Eng Phys 31:1343–1348. https://doi.org/10.1016/j.medengphy.2009.08.006
27. Haddad SMH, Dhaliwal SS, Rotenberg BW, Samani A, Ladak HM (2018) Estimation of the Young's moduli of fresh human oropharyngeal soft tissues using indentation testing. J Mech Behav Biomed Mater 86:352–358. https://doi.org/10.1016/j.jmbbm.2018.07.004
28. Brown EC, Cheng S, McKenzie DK, Butler JE, Gandevia SC, Bilston LE (2015) Tongue stiffness is lower in patients with obstructive sleep apnea during wakefulness compared with matched control subjects. Sleep 38:537–544. https://doi.org/10.5665/sleep.4566
29. Cheng S, Gandevia SC, Green M, Sinkus R, Bilston LE (2011) Viscoelastic properties of the tongue and soft palate using MR elastography. J Biomech 44:450–454. https://doi.org/10.1016/j.jbiomech.2010.09.027
30. Stavness I, Hannam AG, Lloyd JE, Fels S (2008) Towards predicting biomechanical consequences of jaw reconstruction. In: Proceedings of 30th annual international conference of the IEEE engineering in medicine and biology society. EMBS'08—personalized healthcare through technology, IEEE Computer Society, pp 4567–4570. https://doi.org/10.1109/iembs.2008.4650229
31. Gerard JM, Ohayon J, Luboz V, Perrier P, Payan Y (2005) Non-linear elastic properties of the lingual and facial tissues assessed by indentation technique: application to the biomechanics of speech production. Med Eng Phys 27:884–892. https://doi.org/10.1016/j.medengphy.2005.08.001
32. Buchaillard S, Brix M, Perrier P, Payan Y (2007) Simulations of the consequences of tongue surgery on tongue mobility: Implications for speech production in post-surgery conditions. Int J Med Robot Comput Assist Surg 3:252–261. https://doi.org/10.1002/rcs.142
33. Mir M, Ali MN, Ansari U, Sami J (2016) Structure and motility of the esophagus from a mechanical perspective. Esophagus 13:8–16. https://doi.org/10.1007/s10388-015-0497-1
34. Hajhosseini P, Takalloozadeh M (2019) An isotropic hyperelastic model of esophagus tissue layers along with three-dimensional simulation of esophageal peristaltic behavior. J Bioeng Res 1:12–27. https://doi.org/10.22034/jbr.2019.189018.1009
35. Takeda T, Kassab G, Liu J, Puckett JL, Mittal RR, Mittal RK (2002) A novel ultrasound technique to study the biomechanics of the human esophagus in vivo. Am J Physiol Gastrointest Liver Physiol 282. https://doi.org/10.1152/ajpgi.00394.2001

36. Egorov VI, Schastlivtsev IV, Prut EV, Baranov AO, Turusov RA (2002) Mechanical properties of the human gastrointestinal tract. J Biomech 35:1417–1425. https://doi.org/10.1016/S0021-9290(02)00084-2
37. Orvar KB, Gregersen H, Christensen J (1993) Biomechanical characteristics of the human esophagus. Dig Dis Sci 38:197–205. https://doi.org/10.1007/BF01307535
38. Yang J, Liao D, Zhao J, Gregersen H (2004) Shear modulus of elasticity of the esophagus. Ann Biomed Eng 32:1223–1230. https://doi.org/10.1114/B:ABME.0000039356.24821.6c
39. Bou Jawde S, Takahashi A, Bates JHT, Suki B (2020) An analytical model for estimating alveolar wall elastic moduli from lung tissue uniaxial stress-strain curves. Front Physiol 11:121. https://doi.org/10.3389/fphys.2020.00121
40. Andrikakou P, Vickraman K, Arora H (2016) On the behaviour of lung tissue under tension and compression. Sci Rep 6:1–10. https://doi.org/10.1038/srep36642
41. Mercer RR, Crapo JD (1990) Spatial distribution of collagen and elastin fibers in the lungs. J Appl Physiol 69:756–765. https://doi.org/10.1152/jappl.1990.69.2.756
42. Mercer RR, Russell ML, Crapo JD (1994) Alveolar septal structure in different species. J Appl Physiol 77:1060–1066. https://doi.org/10.1152/jappl.1994.77.3.1060
43. Booth AJ, Hadley R, Cornett AM, Dreffs AA, Matthes SA, Tsui JL et al (2012) Acellular normal and fibrotic human lung matrices as a culture system for in vitro investigation. Am J Respir Crit Care Med 186:866–876. https://doi.org/10.1164/rccm.201204-0754OC
44. Han L, Dong H, McClelland JR, Han L, Hawkes DJ, Barratt DC (2017) A hybrid patient-specific biomechanical model based image registration method for the motion estimation of lungs. Med Image Anal 39:87–100. https://doi.org/10.1016/j.media.2017.04.003
45. Concha F, Sarabia-Vallejos M, Hurtado DE (2018) Micromechanical model of lung parenchyma hyperelasticity. J Mech Phys Solids 112:126–144. https://doi.org/10.1016/j.jmps.2017.11.021
46. Ramião NG, Martins PS, Rynkevic R, Fernandes AA, Barroso M, Santos DC (2016) Biomechanical properties of breast tissue, a state-of-the-art review. Biomech Model Mechanobiol 15:1307–1323. https://doi.org/10.1007/s10237-016-0763-8
47. Krouskop TA, Wheeler TM, Kallel F, Garra BS, Hall T (1998) Elastic moduli of breast and prostate tissues under compression. Ultrason Imaging 20:260–274. https://doi.org/10.1177/016173469802000403
48. Sutradhar A, Miller MJ (2013) In vivo measurement of breast skin elasticity and breast skin thickness. Ski Res Technol 19:e191–e199. https://doi.org/10.1111/j.1600-0846.2012.00627.x
49. Ulger H, Erdogan N, Kumanlioglu S, Unur E (2003) Effect of age, breast size, menopausal and hormonal status on mammographic skin thickness. Ski Res Technol 9:284–289. https://doi.org/10.1034/j.1600-0846.2003.00027.x
50. Huang SY, Boone JM, Yang K, Kwan ALC, Packard NJ (2008) The effect of skin thickness determined using breast CT on mammographic dosimetry. Med Phys 35:1199–1206. https://doi.org/10.1118/1.2841938
51. Gold RH, Montgomery CK, Minagi H, Annes GP (1971) The significance of mammary skin thickening in disorders other than primary carcinoma: a roentgenologic-pathologic correlation. Am J Roentgenol Radium Ther Nucl Med 112:613–621. https://doi.org/10.2214/ajr.112.3.613
52. Willson SA, Adam EJ, Tucker AK (1982) Patterns of breast skin thickness in normal mammograms. Clin Radiol 33:691–693. https://doi.org/10.1016/S0009-9260(82)80407-8
53. Samani A, Zubovits J, Plewes D (2007) Elastic moduli of normal and pathological human breast tissues: an inversion-technique-based investigation of 169 samples. Phys Med Biol 52:1565–1576. https://doi.org/10.1088/0031-9155/52/6/002
54. Cox TR, Erler JT (2011) Remodeling and homeostasis of the extracellular matrix: implications for fibrotic diseases and cancer. DMM Dis Model Mech 4:165–178. https://doi.org/10.1242/dmm.004077
55. Unlu MZ, Krol A, Magri A, Mandel JA, Lee W, Baum KG et al (2010) Computerized method for nonrigid MR-to-PET breast-image registration. Comput Biol Med 40:37–53. https://doi.org/10.1016/j.compbiomed.2009.10.010
56. Gefen A, Dilmoney B (2007) Mechanics of the normal woman's breast. Technol Heal Care 15:259–271. https://doi.org/10.3233/thc-2007-15404

57. Van Houten EEW, Doyley MM, Kennedy FE, Weaver JB, Paulsen KD (2003) Initial in vivo experience with steady-state subzone-based MR elastography of the human breast. J Magn Reson Imaging 17:72–85. https://doi.org/10.1002/jmri.10232
58. Arroyo J, Saavedra AC, Guerrero J, Montenegro P, Aguilar J, Pinto JA et al (2018) Breast elastography: identification of benign and malignant cancer based on absolute elastic modulus measurement using vibro-elastography. In: Proceedings of SPIE Image perception, observer, performance, and technology assessment, vol 10577, pp 48. https://doi.org/10.1117/12.2293664
59. Brandstaeter S, Fuchs SL, Aydin RC, Cyron CJ (2019) Mechanics of the stomach: a review of an emerging field of biomechanics. GAMM-Mitteilungen 42. https://doi.org/10.1002/gamm.201900001
60. Aydin RC, Brandstaeter S, Braeu FA, Steigenberger M, Marcus RP, Nikolaou K et al (2017) Experimental characterization of the biaxial mechanical properties of porcine gastric tissue. J Mech Behav Biomed Mater 74:499–506. https://doi.org/10.1016/j.jmbbm.2017.07.028
61. Zhao J, Liao D, Gregersen H (2005) Tension and stress in the rat and rabbit stomach are location- and direction-dependent. Neurogastroenterol Motil 17:388–398. https://doi.org/10.1111/j.1365-2982.2004.00635.x
62. Zhao J, Liao D, Chen P, Kunwald P, Gregersen H (2008) Stomach stress and strain depend on location, direction and the layered structure. J Biomech 41:3441–3447. https://doi.org/10.1016/j.jbiomech.2008.09.008
63. Rosen J, Brown JD, De S, Sinanan M, Hannaford B (2008) Biomechanical properties of abdominal organs in vivo and postmortem under compression loads. J Biomech Eng 130. https://doi.org/10.1115/1.2898712
64. Lim YJ, Deo D, Singh TP, Jones DB, De S (2009) In situ measurement and modeling of biomechanical response of human cadaveric soft tissues for physics-based surgical simulation. Surg Endosc 23:1298–1307. https://doi.org/10.1007/s00464-008-0154-z
65. Kemper AR, Santago AC, Stitzel JD, Sparks JL, Duma SM (2012) Biomechanical response of human spleen in tensile loading. J Biomech 45:348–355. https://doi.org/10.1016/j.jbiomech.2011.10.022
66. Umale S, Deck C, Bourdet N, Dhumane P, Soler L, Marescaux J et al (2013) Experimental mechanical characterization of abdominal organs: liver, kidney & spleen. J Mech Behav Biomed Mater 17:22–33. https://doi.org/10.1016/j.jmbbm.2012.07.010
67. Stingl J, Báĉa V, Ĉech P, Kovanda J, Kovandová H, Mandys V et al (2002) Morphology and some biomechanical properties of human liver and spleen. Surg Radiol Anat 24:285–289. https://doi.org/10.1007/s00276-002-0054-1
68. Galford JE, McElhaney JH (1970) A viscoelastic study of scalp, brain, and dura. J Biomech 3:211–221. https://doi.org/10.1016/0021-9290(70)90007-2
69. Polio SR, Kundu AN, Dougan CE, Birch NP, Aurian-Blajeni DE, Schiffman JD et al (2018) Cross-platform mechanical characterization of lung tissue. PLoS ONE 13:e0204765. https://doi.org/10.1371/journal.pone.0204765

Chapter 5
Tissues in Functional Organs—Medium Stiffness

5.1 Liver

The liver, located on the right side of the abdomen and weighing around 3 pounds, is the second largest organ after the skin. The liver is one of the visceral organs of the human body and performs a variety of important and complicated functions, such as the metabolism of proteins, carbohydrates, and lipids and the storage of minerals and vitamins. Most important, liver is responsible for the production of bile and its absorption into the small intestine, which helps for the effective digestion of meals. The liver tissue is made up of numerous endothelial, vascular, and parenchymal cells (epithelial cells, stellate cells, hepatocytes, and fibroblasts) [1, 2].

To date, porcine and bovine animal models have been used to estimate the mechanical characteristics of liver tissue [3–6] (Fig. 5.1). The liver tissue is one of the most common organs affected during the trauma injuries, which makes the mechanical properties of the liver tissue are of great interest. Considering this fact, the experimental data of the cadaveric liver tissue is anticipated for the analysis of trauma studies and injury biomechanics under the realistic mechanical loading conditions [7]. The literature studies reported the average elastic modulus of liver tissue under tensile and compressive loading conditions as 10.5 kPa and 90 kPa, respectively [8, 9]. Stingl et al. [10] performed an experimental study using fresh tissue samples from cadavers to determine the mechanical properties of the human liver. Under compressive loading, the estimated elastic modulus range was 66–386 kPa in males and 89–308 kPa in females, with an ultimate stress value of 203 kPa for the tissue. Recent research by Karimi and Shojaei [7] evaluated the axial and transverse mechanical properties of human liver tissue under tensile and compressive loading conditions (Fig. 5.2). Under tension, the elastic modulus for axial loading was 12.16 ± 1.20 kPa, and for transverse loading, it was 7.17 ± 0.85 kPa. The compressive elastic modulus was determined to be 196.54 ± 13.15 kPa for axial loading and 112.41 ± 8.98 kPa for transverse loading. For compression loadings, the assessed ultimate stresses were

A. Chanda and G. Singh, *Mechanical Properties of Human Tissues*,
Materials Horizons: From Nature to Nanomaterials,
https://doi.org/10.1007/978-981-99-2225-3_5

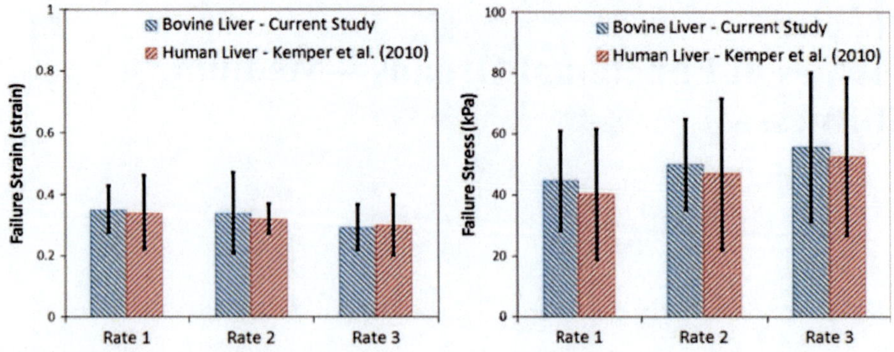

Fig. 5.1 Comparison of tensile failure strain between bovine and human livers [6]

314.01 ± 19.10 kPa (axial) and 355.10 ± 20.10 kPa (transverse), and for tensile loadings, 9.65 ± 1.10 kPa (axial) and 9.52 ± 0.80 kPa (transverse), respectively. Using MR elastography, the average elastic modulus of human liver tissue was determined in vivo and found to be 2.24 ± 0.23 kPa [11].

Numerous researchers have evaluated the elastic modulus of liver tissue using different low and high strain rates. Yeh et al. [12] determined the elastic modulus of human liver tissue to be 0.65 kPa at strain rates of 5%, 1.1 kPa at strain rates of 10%, and 2 kPa at strain rates of 15%. Hollenstein et al. [5] recorded the elastic modulus of the liver capsule with low and high stresses to be 38.5 ± 4.9 MPa and 1.1 ± 0.2 MPa, respectively. Umale et al. [13] estimated the elastic modulus of the liver tissue and reported to be 48.15 ± 4.5 MPa and 8.22 ± 3.42 MPa for the low and high strains, respectively. Brunon et al. [3] reported an average ultimate stress and strain for the human liver tissue. The estimated values were 1.85 ± 1.18 MPa for the ultimate stress with a strain of 32.6 ± 13.8% and mean elastic modulus of 270 kPa [14].

5.2 Gallbladder

The gallbladder tissue, a thin pear-shaped organ, is connected directly beneath the right lobe of the liver. The primary function of the gallbladder tissue is to store and release bile into the duodenum so that the lipids extracted from the food can be utilized for the proper digestion of the meal [15]. Karimi et al. [16] assessed the mechanical characteristics of human gallbladder tissue using a range of axial and transverse loadings (Fig. 5.3). Sixteen-male cadaveric tissue samples were excised and tested a strain rate of 5 mm/min. For the axial loading conditions, the elastic modulus and ultimate stress were found to be 641.20 ± 28.12 kPa and 1240 ± 99.94 kPa, respectively. For the transverse loading conditions, the elastic modulus and ultimate stress were found to be 255 ± 24.55 kPa and 348 ± 66.75 kPa, respectively (Fig. 5.4).

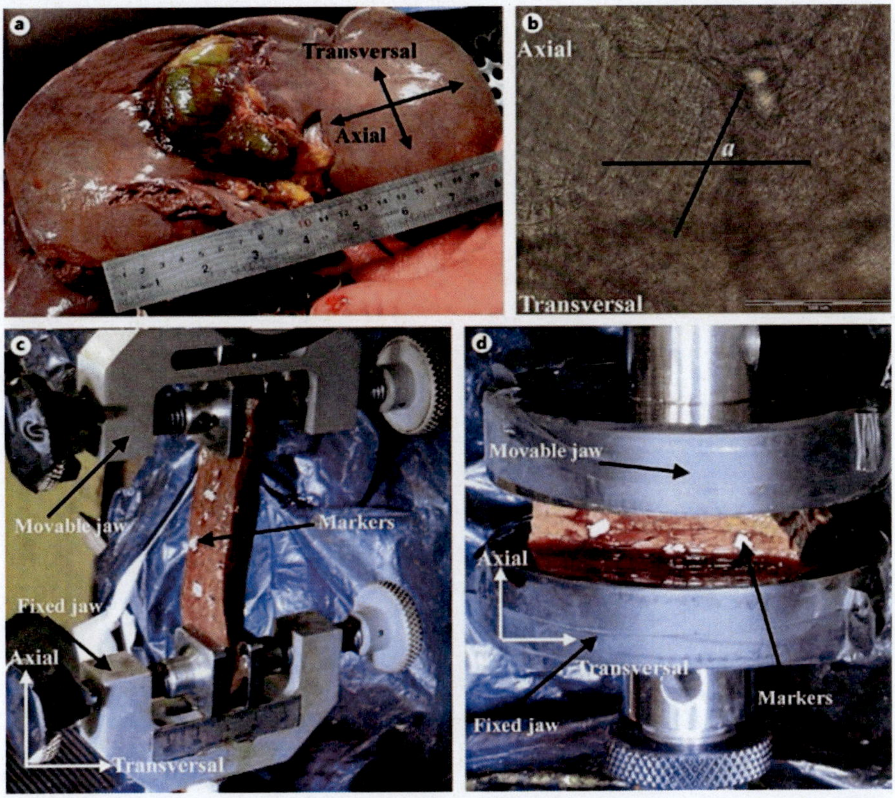

Fig. 5.2 **a** Fresh human cadaveric liver tissues; **b** 3 μm thick histological section with the axial and transversal orientations was prepared; **c** tensile test as well as the **d** compressive test were carried out, and the markers were tracked using the CCD cameras [7]

The work carried out by Karimi et al. [16] is the sole research to establish the tensile characteristics of the human gallbladder under complex loading circumstances.

5.3 Kidney

The kidneys are a pair of bean-shaped organs positioned along the posterior wall of the abdominal cavity. As the right side of the liver is much larger than the left side, the left kidney is positioned somewhat above the right kidney. The kidneys are in contact with the back muscles and are located posterior to the peritoneum. The kidneys are protected by a thin layer of adipose, which keeps them in place and prevents physical damage. The function of kidneys is to filter metabolic wastes, excess ions, and chemicals from the blood, maintain the water balance of water, salts, and minerals, and are almost twice stiff as the liver [17].

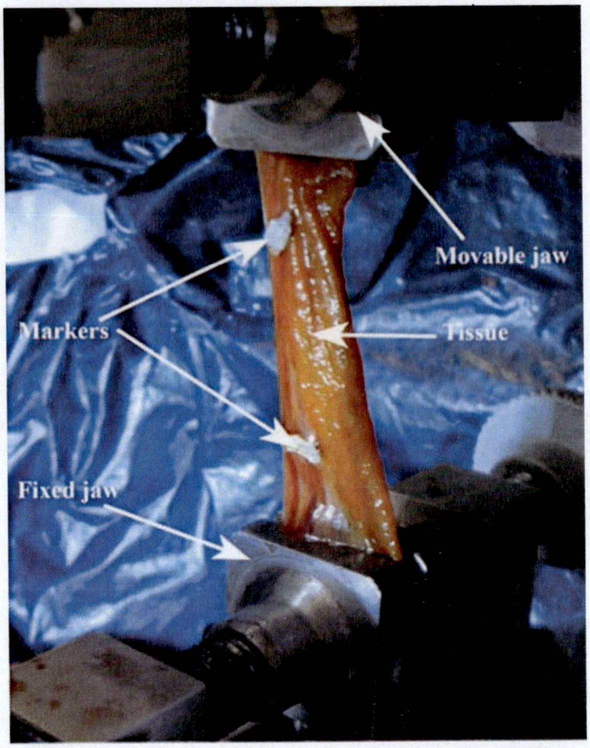

Fig. 5.3 Tensile testing of the human cadaveric gallbladder tissue [16]

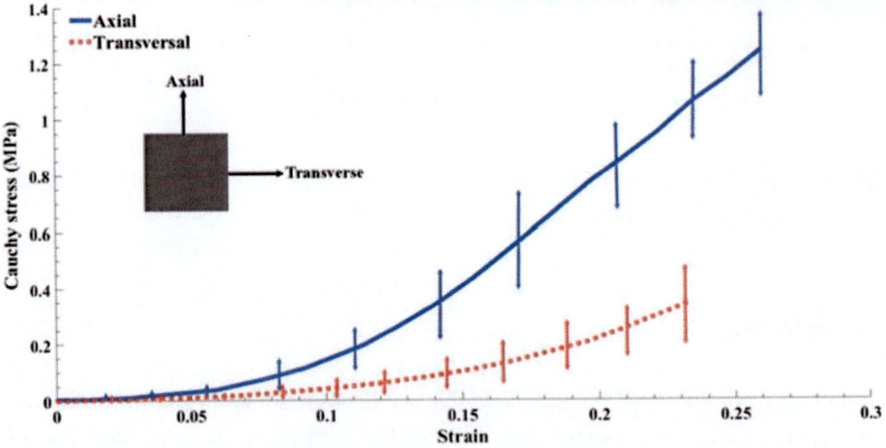

Fig. 5.4 Uniaxial tensile stress–strain of the gallbladder tissues under the axial and transversal loadings [16]

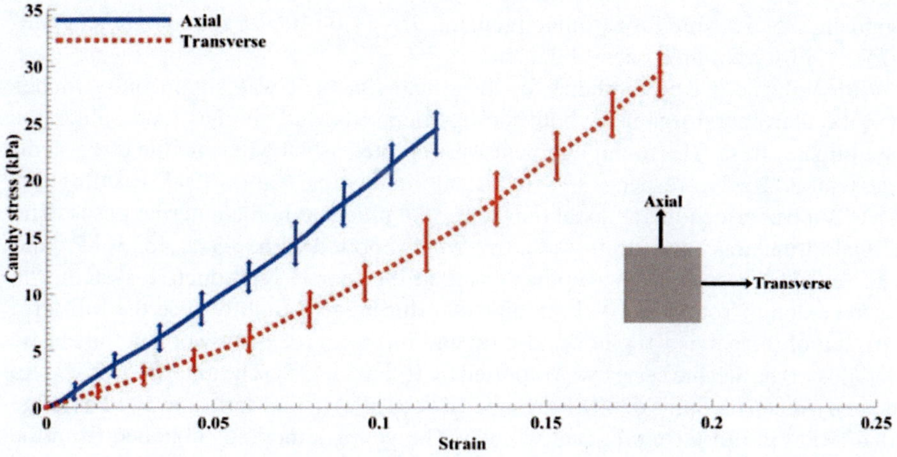

Fig. 5.5 Uniaxial tensile stress–strain of the kidney tissue under the axial and transversal loadings [19]

Snedeker et al. [18] estimated the mechanical properties of the kidney capsular membrane and reported the values as 41.5 MPa, 9.0 ± 2.9 MPa, and 33.4 ± 6.5% for elastic modulus, ultimate stress, and strain, respectively. Karimi and Shojaei [19] evaluated the mechanical properties of the human kidney under axial and transverse tensile loads. The elastic modulus and ultimate stresses were recorded to be 180.32 kPa and 24.46 kPa for axial loading and 95.64 kPa and 31 kPa for transversal tensile loading, respectively (Fig. 5.5).

5.4 Uterus

The uterus connects to the vagina and fallopian tubes through the cervix. This tissue is typically pear-shaped and is around 7.6 cm in length, 4.5 cm in width, and 3.0 cm in thickness [20, 21]. To accommodate the fetus, the uterus tissue shows significant deformation and changes its mechanical properties during the gestational growth. The muscles of the uterus show isotropic behavior at the early stage of labor and then the anisotropic behavior as the labor progress [22]. The uterus is surrounded by endopelvic fascial ligaments, namely pubocervical ligaments, cardinal ligaments, uterosacral ligaments, and round ligaments. The uterus, often known as the womb, is mostly composed of soft muscular tissue, collagen, and elastin fibers [21]. The elastic modulus of the uterus tissue was reported to be 5 kPa with a Poisson's ratio of 0.49 [23]. A study by Baah-Dwomoh et al. [21] quantified the modulus of elasticity of the uterus tissue and also for the different ligaments of the uterus tissue. The estimated elastic modulus values were 0.02–1.4 MPa for uterus tissue, 2.17–243 kPa for cervix ligament, 0.5–5.4 MPa for cardinal ligament, 9.1–14.0 MPa for round ligament, and 0.75–29.8 MPa for uterosacral ligament.

Although the elastic modulus for the round ligament was significantly higher than the uterosacral ligament, both the ligaments showed similar trend values for the ultimate stress. The round ligament was reported as 4.1 MPa and the uterosacral ligament as 4 MPa, respectively, for the ultimate stress values [21]. Bisplinghoff et al. [24] performed the uniaxial tensile for the pregnant human uterine tissue. The ultimate true stress and strain at failure were reported as 656.3 ± 483.9 kPa and 0.32 ± 0.112, respectively. Another tissue in the female reproductive system, the cervix, extends from 1 to 10 cm in diameter during gestation to force the full-term fetus out of the pelvis [21]. For mid-term and full-term pregnant women, the elastic modulus of the uterine cervix was reported by Hee et al. [25]. The anterior cervical lip showed a mean modulus of 80 kPa (range 50–140 kPa) in mid-term and 30 kPa (range 20–40 kPa) in full-term pregnant women. The young's modulus obtained from the posterior cervical lip was 90 kPa (range 60–150 kPa) in mid-term and 70 kPa (range 30–120 kPa) in full-term pregnant women. The knowledge of the realistic mechanical properties can be used for developing computational model for investigating various disease and injury biomechanics scenarios (Fig. 5.6).

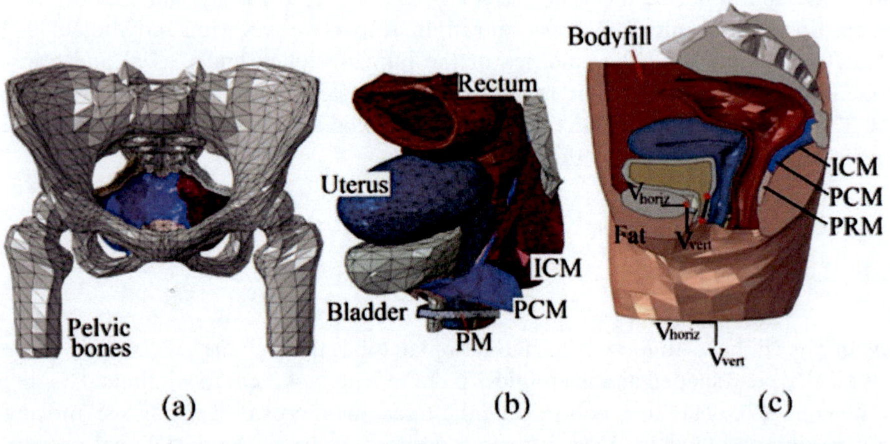

Fig. 5.6 **a** Front view of the pelvic floor muscles and pelvic bone, **b** anterolateral view of the pelvic floor organs and muscles, and **c** mid-sagittal view of the complete pelvic model. Abbreviation used in this figure: ICM–Iliococcygeus muscle, PCM–pubococcygeus muscle, PRM–puborectalis muscle, and PM–perineal membrane [24]

5.5 Summarizing Mechanical Properties of Tissues with Medium Stiffness

Table 5.1 outlines the mechanical properties of human soft tissues in functional organs with medium stiffness. The values of Table 5.1 were used to plot a graph, and Fig. 5.7 illustrates the modulus of elasticity (mean ± S.D.) of the discussed soft tissues.

Table 5.1 Mechanical properties of the medium stiff tissues in functional organs

Tissue	Reference	Modulus of elasticity	Ultimate tensile strength	Subgroup/remarks
Liver	Evans et al. [8]	10.5 kPa	–	Tensile
	Nava et al. [9]	90 kPa	–	Compressive
	Stingl et al. [10]	66–386 kPa (male) 89–308 kPa (female)	203 kPa	Compressive
	Karimi and Shojaei [7]	12.16 ± 1.20 kPa	9.65 ± 1.10 kPa	Tensile
	Karimi and Shojaei [7]	196.54 ± 13.15 kPa	314.01 ± 19.10 kPa	Compressive
	Brunon et al. [3]	–	1.85 ± 1.18 MPa	–
	Mattei and Ahluwalia [14]	270 kPa	–	–
Gallbladder	Karimi et al. [16]	641.20 ± 28.12 kPa	1240 ± 99.94 kPa	Axial loading
	Karimi et al. [16]	255 ± 24.55 kPa	348 ± 66.75 kPa	Transversal loading
Kidney	Snedeker et al. [18]	41.5 MPa	9.0 ± 2.9 MPa	Capsular membrane
	Karimi and Shojaei [19]	180.32 kPa	24.46 kPa	Axial direction
	Karimi and Shojaei [19]	95.64 kPa	31 kPa	Transversal directions
Uterus	Baah-Dwomoh et al. [21]	0.02–1.4 MPa	–	–
	Bisplinghoff et al. [24]	656.3 ± 483.9 kPa	–	–
	Bisplinghoff et al. [21]	2.17–243 kPa	–	Cervix
	Hee et al. [25]	80 kPa	–	anterior Cervical lip (mid-term pregnancy)

(continued)

Table 5.1 (continued)

Tissue	Reference	Modulus of elasticity	Ultimate tensile strength	Subgroup/remarks
	Hee et al. [25]	30 kPa	–	Anterior cervical lip (full-term pregnancy)
	Hee et al. [25]	90 kPa	–	Posterior cervical lip (mid-term pregnancy)
	Hee et al. [25]	70 kPa	–	Posterior cervical lip (full-term pregnancy)
	Bisplinghoff et al. [21]	0.5–5.4 MPa	–	Cardinal ligament
	Bisplinghoff et al. [21]	9.1–14.0 MPa	4.1 MPa	Round ligament
	Bisplinghoff et al. [21]	0.75–29.8 MPa	4 MPa	Uterosacral ligament

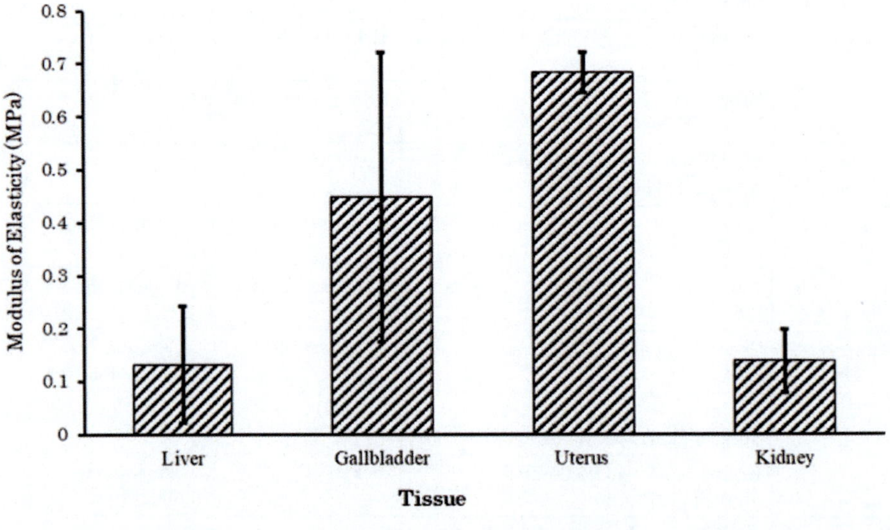

Fig. 5.7 Modulus of elasticity (mean ± S.D.) of the medium stiff tissues in functional organs

References

1. Kazemnejad S (2009) Hepatic tissue engineering using scaffolds: state of the art. Avicenna J Med Biotechnol 1:135–145
2. Lorenzini S, Andreone P (2007) Stem cell therapy for human liver cirrhosis: a cautious analysis of the results. Stem Cells 25:2383–2384. https://doi.org/10.1634/stemcells.2007-0056

3. Brunon A, Bruyère-Garnier K, Coret M (2010) Mechanical characterization of liver capsule through uniaxial quasi-static tensile tests until failure. J Biomech 43:2221–2227. https://doi.org/10.1016/j.jbiomech.2010.03.038
4. Chui C, Kobayashi E, Chen X, Hisada T, Sakuma I (2007) Transversely isotropic properties of porcine liver tissue: experiments and constitutive modelling. Med Biol Eng Comput 45:99–106. https://doi.org/10.1007/s11517-006-0137-y
5. Hollenstein M, Nava A, Valtorta D, Snedeker JG, Mazza E. Mechanical characterization of the liver capsule and parenchyma. Lect Notes Comput Sci (Incl Subser Lect Notes Artif Intell Lect Notes Bioinform) 4072:150–158 (LNCS, Springer). https://doi.org/10.1007/11790273_17
6. Lu YC, Kemper AR, Untaroiu CD (2014) Effect of storage on tensile material properties of bovine liver. J Mech Behav Biomed Mater 29:339–349. https://doi.org/10.1016/j.jmbbm.2013.09.022
7. Karimi A, Shojaei A (2018) An experimental study to measure the mechanical properties of the human liver. Dig Dis 36:150–155. https://doi.org/10.1159/000481344
8. Evans DW, Moran EC, Baptista PM, Soker S, Sparks JL (2013) Scale-dependent mechanical properties of native and decellularized liver tissue. Biomech Model Mechanobiol 12:569–580. https://doi.org/10.1007/s10237-012-0426-3
9. Nava A, Mazza E, Kleinermann F, Avis NJ, McClure J, Bajka M (2004) Evaluation of the mechanical properties of human liver and kidney through aspiration experiments. Technol Heal Care 12:269–280. https://doi.org/10.3233/thc-2004-12306
10. Stingl J, Báĉa V, Ĉech P, Kovanda J, Kovandová H, Mandys V et al (2002) Morphology and some biomechanical properties of human liver and spleen. Surg Radiol Anat 24:285–289. https://doi.org/10.1007/s00276-002-0054-1
11. Huwart L, Peeters F, Sinkus R, Annet L, Salameh N, ter Beek LC et al (2006) Liver fibrosis: non-invasive assessment with MR elastography. NMR Biomed 19:173–179. https://doi.org/10.1002/nbm.1030
12. Yeh WC, Li PC, Jeng YM, Hsu HC, Kuo PL, Li ML et al (2002) Elastic modulus measurements of human liver and correlation with pathology. Ultrasound Med Biol 28:467–474. https://doi.org/10.1016/S0301-5629(02)00489-1
13. Umale S, Chatelin S, Bourdet N, Deck C, Diana M, Dhumane P et al (2011) Experimental in vitro mechanical characterization of porcine Glisson's capsule and hepatic veins. J Biomech 44:1678–1683. https://doi.org/10.1016/j.jbiomech.2011.03.029
14. Mattei G, Ahluwalia A (2016) Sample, testing and analysis variables affecting liver mechanical properties: a review. Acta Biomater 45:60–71. https://doi.org/10.1016/j.actbio.2016.08.055
15. Li WG, Hill NA, Ogden RW, Smythe A, Majeed AW, Bird N et al (2013) Anisotropic behaviour of human gallbladder walls. J Mech Behav Biomed Mater 20:363–375. https://doi.org/10.1016/j.jmbbm.2013.02.015
16. Karimi A, Shojaei A, Tehrani P (2017) Measurement of the mechanical properties of the human gallbladder. J Med Eng Technol 41:541–545. https://doi.org/10.1080/03091902.2017.1366561
17. Nicolle S, Palierne JF (2010) Dehydration effect on the mechanical behaviour of biological soft tissues: observations on kidney tissues. J Mech Behav Biomed Mater 3:630–635. https://doi.org/10.1016/j.jmbbm.2010.07.010
18. Snedeker JG, Niederer P, Schmidlin FR, Farshad M, Demetropoulos CK, Lee JB et al (2005) Strain-rate dependent material properties of the porcine and human kidney capsule. J Biomech 38:1011–1021. https://doi.org/10.1016/j.jbiomech.2004.05.036
19. Karimi A, Shojaei A (2017) Measurement of the mechanical properties of the human kidney. IRBM 38:292–297. https://doi.org/10.1016/j.irbm.2017.08.001
20. Li X, Kruger JA, Chung J-H, Nash MP, Nielsen PMF (2008) Modelling childbirth: comparing athlete and non-athlete pelvic floor mechanics. Springer, Berlin, Heidelberg, pp 750–757. https://doi.org/10.1007/978-3-540-85990-1_90
21. Baah-Dwomoh A, McGuire J, Tan T, De Vita R (2016) Mechanical properties of female reproductive organs and supporting connective tissues: a review of the current state of knowledge. Appl Mech Rev 68. https://doi.org/10.1115/1.4034442

22. Mizrahi J, Karni Z, Polishuk WZ (1980) Isotropy and anisotropy of uterine muscle during labor contraction. J Biomech 13:211–218. https://doi.org/10.1016/0021-9290(80)90364-4
23. Dias N, Peng Y, Khavari R, Nakib NA, Sweet RM, Timm GW et al (2017) Pelvic floor dynamics during high-impact athletic activities: a computational modeling study. Clin Biomech 41:20–27. https://doi.org/10.1016/j.clinbiomech.2016.11.003
24. Bisplinghoff JA, Kemper AR, Duma SM (2012) Dynamic material properties of the pregnant human uterus. J Biomech 45:1724–1727. https://doi.org/10.1016/j.jbiomech.2012.04.001
25. Hee L, Sandager P, Petersen O, Uldbjerg N (2013) Quantitative sonoelastography of the uterine cervix by interposition of a synthetic reference material. Acta Obstet Gynecol Scand 92:1244–1249. https://doi.org/10.1111/aogs.12246

Chapter 6
Tissues in Functional Organs—High Stiffness

6.1 Nasal Cavity

The nasal cavity is the first airway pass and is the primary external opening of the respiratory system. The region of the anterior nasal cavity is supported and protected by the bone, cartilage, muscle, and skin of the nose (Fig. 6.1). The function of the nasal cavity is to warm, filter, and moisten the air before it reaches the lungs. The nasal cavity is lined with hairs and mucus to prevent dust, pollen, etc., from entering the body during the inhale process.

To date, no significant work has been reported to quantity the mechanical characteristics of the nasal tissue. Rotter et al. [2] performed an experimental study on the human nasal cartilage and estimated its stiffness with an elastic modulus of 0.23 ± 0.02 MPa. Richmon et al. [3] recorded an equilibrium modulus of 0.44 ± 0.04 MPa for the compressive analysis of human nasal septum. In addition, the human nasal septal cartilages were reported to have an equilibrium modulus of 3.01 ± 0.39 MPa, a dynamic modulus of 4.99 ± 0.44 MPa, an ultimate stress of 1.90 ± 0.24 MPa, and a failure strain of 0.35 ± 0.03 mm/mm during tensile testing [4]. However, an experimental investigation by Alkan et al. [5] reported that the average elastic modulus of the 18 fresh cadaveric septal cartilages was recorded to be 1.39 MPa. Griffin et al. [6] determined the mean elastic modulus of the septal cartilage to be 2.72 ± 0.63 MPa and 2.09 ± 0.81 MPa for the medial and lateral alar cartilages, respectively. The biomechanical properties of the nasal periosteum and fascia were estimated by Zeng et al. [7] under uniaxial tensile loading conditions. Zeng et al. [6] examined the biomechanical characteristics of the human nasal periosteum and fascia during uniaxial elongation. The mean tensile strength of periosteum was 3.88 MPa with a failure strain of 24.8%, whereas the mean tensile strength of fascia samples was 2.70 MPa with a failure strain of 51.8%.

A. Chanda and G. Singh, *Mechanical Properties of Human Tissues*,
Materials Horizons: From Nature to Nanomaterials,
https://doi.org/10.1007/978-981-99-2225-3_6

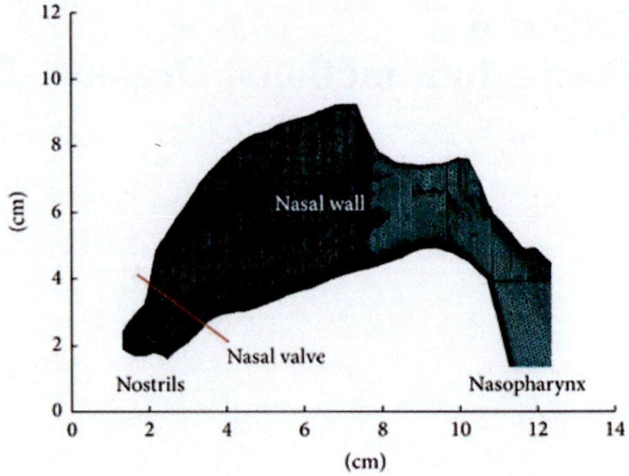

Fig. 6.1 Three-dimensional finite element model of a nasal cavity [1]

6.2 Oral Cavity

The oral cavity or the mouth is a secondary opening/path to the respiratory system, in addition to the nasal cavity. The air pathway via oral cavity is shorter than that through the nose, but it does not filtrate, warm, and moisturize the air efficiently due to the absence of hairs and sticky mucus. However, the oral cavity is advantageous in some cases as it has larger diameter and shorter pathway than the nasal cavity, which facilitates the fast reaching of air to the lungs.

To date, the literature on the mechanical properties of the human oral cavity limited. In a recent study by Choi et al. [8], they estimated the elastic modulus and ultimate tensile stress for the oral mucosa tissues of human (Fig. 6.2). In the gingiva group, the maximum elastic modulus was recorded to be 37.36 ± 17.4 MPa, followed by the hard palate with elastic modulus of 18.13 ± 4.51 MPa and 8.36 ± 5.86 MPa for buccal mucosa. The ultimate tensile strength of the tissue samples varied from 1.54 ± 0.5 MPa (buccal mucosa) to 3.81 ± 0.9 MPa (gingiva).

Indentation technique was used by Haddad et al. [9] to determine the elastic modulus for the uvula and soft palate and recorded their respective elastic modulus as 2.63 ± 1.08 kPa and 6.54 ± 3.52 kPa. A similar study was reported by Malhotra et al., and they estimated an average elastic modulus of 6 kPa for the uvula and soft palate [10]. Birch and Srodon [11] conducted uniaxial tensile testing at different locations of the soft palate and found that a range for the elastic modulus, i.e., 0.585 kPa–1.410 kPa. The average Poisson ratio in the inferior-superior direction was 0.45 ± 0.26 and 0.30 ± 0.21 in the lateral direction. Magnetic resonance elastography was used by Brown et al. [12] to record the shear modulus of the soft palate, and the measure value was 2.42 kPa.

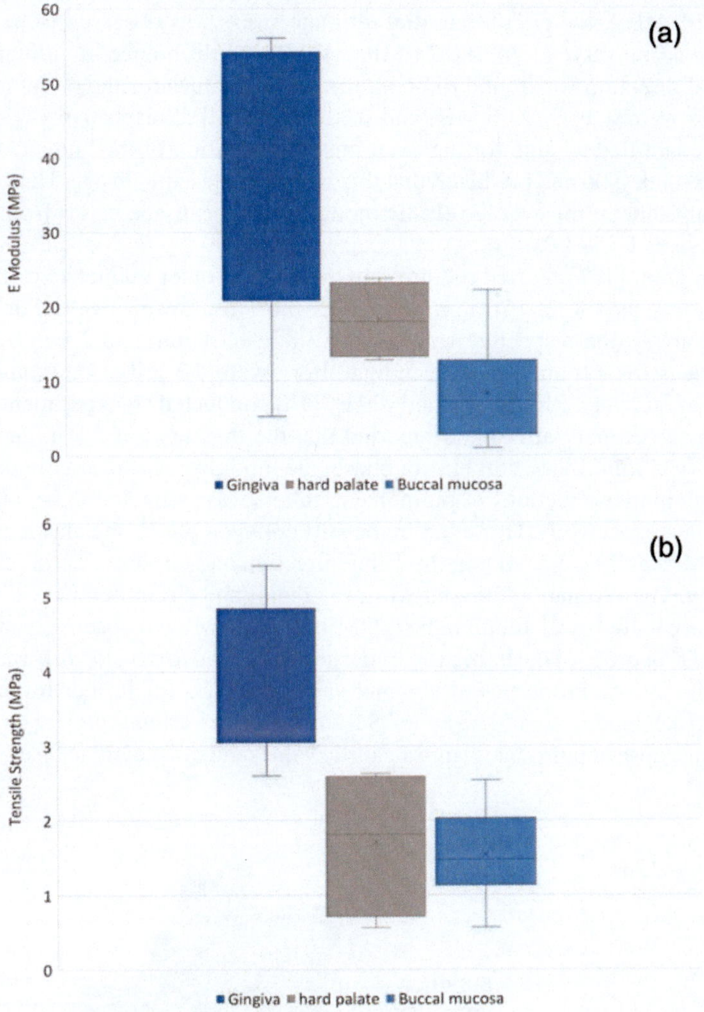

Fig. 6.2 Graph showing **a** elastic modulus and **b** tensile strength of the oral cavity tissues [8]

6.3 Heart

The heart is a pumping organ located in the thoracic area, which is medial to the lungs and to the body's midline. The human heart is 1/3 on the right side and remaining 2/3 on the left. At a stress level of 1.0 MPa [13], it was observed that the pulmonary valve elastic modulus (16.05 ± 2.02 MPa) and aortic valve elastic modulus (15.34 ± 3.82 MPa) exhibited similar values in the circumferential direction. In the radial direction, the pulmonary valve cusps have a lower elastic modulus than the aortic valve cusps (1.32 ± 0.93 MPa vs. 1.98 ± 0.15 MPa). For the pulmonary valve cusps

(2.78 ± 1.05 MPa), the circumferential ultimate stress was observed to be greater than for the aortic valve (1.74 ± 0.24 MPa). While the difference in ultimate stress in the radial direction was found to be minimum, the values for the pulmonary and aortic cusps were 0.29 ± 0.06 MPa and 0.32 ± 0.04 MPa, respectively. Holzapfel et al. [14] quantified a value for the axial in situ elongation of the human coronary artery as 1.044 ± 0.06 and predicted that this value decreased with age. Under varied stress circumstances, the average elastic modulus of heart tissue varies from 1.27 ± 0.63 to 27.90 ± 10.59 kPa.

Canham et al. [15] reported the proximal and distal outer diameters of the right coronary artery as 4.4 ± 0.76 mm and 3.2 ± 0.67 mm, respectively. For the left circumflex artery, the respective values were 3.8 ± 0.61 mm and 2.5 ± 0.74 mm, and the values for left anterior descending artery were 4.3 ± 0.47 mm and 2.5 ± 0.33 mm, respectively. Richardson and Keeny [16] conducted an experimental study on the human coronary artery and reported that the thickness of the intima ranges from 0.15–0.35 mm. In the soft elastin phase, the modulus of elasticity in the radial and circumferential directions of pulmonary leaflets was estimated to be ~0.3 MPa and ~1 MPa, respectively. However, in the stiff collagen phase, the elastic modulus was ~3.8 MPa and ~15.5 MPa in the radial direction and circumferential direction, respectively. The ultimate failure stress in the radial and circumferential directions of pulmonary leaflets was found to be 0.5 MPa and 1.5 MPa, respectively [17]. The mechanical properties of the human heart tissue anticipated to develop full-scale heart model with realistic properties to conduct various studies for further investigation (Fig. 6.3). Such models would be beneficial for planning various surgical treatments including prolapse of pulmonary, mitral, aortic value, and other cardiac diseases (e.g., stenosis).

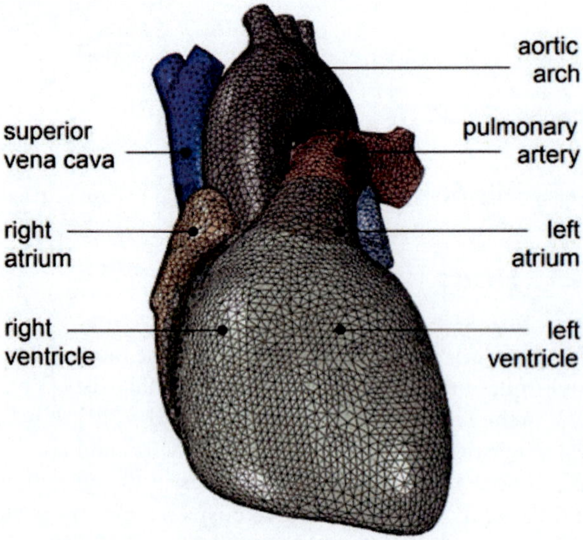

Fig. 6.3 Finite element model of the human heart [18]

6.4 Pancreas

The pancreas is a large organ situated just inferior and posterior to the stomach in the abdominal cavity. With its "head" linked to the duodenum and its "tail" pointing to the left wall of the abdominal cavity, it is about 6 in. long and shaped like a small and lumpy snake. For the chemical breakdown of carbohydrates, the pancreas secretes digestive enzymes into the small intestine. As it comprises both endocrine and exocrine glandular tissue, the pancreas is a heterocrine gland. The mean shear stiffness for the pancreas tissue was reported using in vivo measurement by Shi et al. [19] with magnetic resonance elastography technique. The observed values were 1.2 kPa at 40 Hz and 2.1 kPa at 60 Hz, respectively. The pancreas is more viscous than other abdominal organs such as the kidney, liver, and spleen [20]. To date, the mechanical properties of the human pancreas tissue are not significantly characterized by the researchers.

6.5 Small Intestine

The small intestine is a long and thin tube with a diameter of around 1 in. and a length of 20–25 ft. It is part of the lower gastrointestinal tract, which also includes the duodenum, jejunum, and ileum [21]. It is positioned under the stomach and occupies the majority of the abdominal cavity. The whole small intestine is coiled like a hose, and its inner surface is covered with many ridges and folds. These folds promote digestion and absorption of nutrients [22].

Under macroscopic mechanical behavior, the intestinal tissue exhibited a nonlinear stress–strain relationship, anisotropy, and viscoelasticity [23]. Egorov et al. [24] tested small intestine tissue for the uniaxial tensile and reported an ultimate stress value of 0.9 MPa with a failure strain of 140% (Fig. 6.4). However, to describe the anisotropic response of the intestinal wall, no adequate details were reported. The intestinal samples were excised from four fresh human intestines and four embalmed cadaveric intestines by Bourgouin et al. [25] to investigate the tensile properties at 1 m/s. The elastic modulus for the primary phase (outer layer response), i.e., the pre-damage structure, was 5.16 ± 3.03 MPa for embalmed samples and 2.69 ± 0.37 MPa for fresh intestinal samples. The secondary process (inner layer response) was correlated to the damaged structure, and evaluated elastic modulus was 1.61 ± 0.43 MPa for embalmed cadaveric samples and 1.25 ± 0.07 MPa for the fresh human intestines.

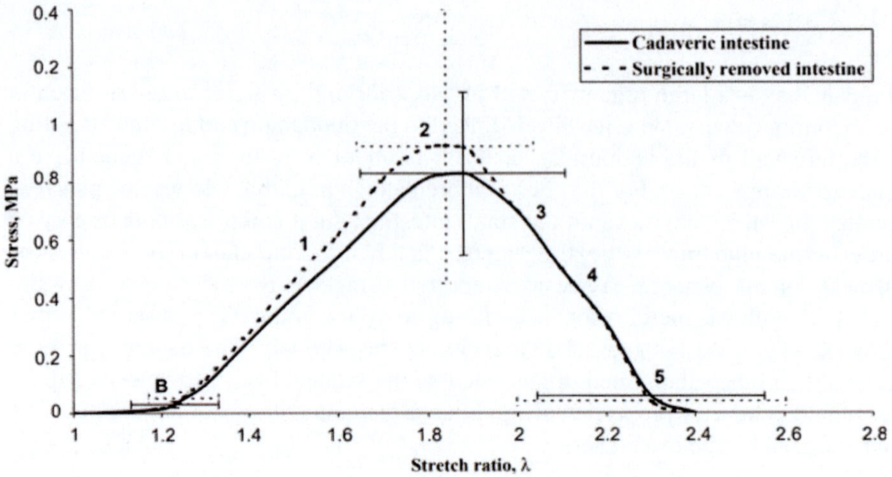

Fig. 6.4 Mechanical testing of human cadaveric and surgically removed transversal specimens of small intestine [24]

6.6 Colon

The colon, also known as the large intestine, is a tube around 2.5 in. in diameter and 1.5 m in length. It is slightly inferior to the stomach and coils along the top and lateral edge of the small intestine. Massalou et al. [26] experimentally tested the human colon tissue under uniaxial tensile loading conditions and varying loading circumstances, such as dynamic loading, intermediate loading, and quasi-static loading. The longitudinal sample exhibited an elastic modulus of 3.17 $\pm$ 2.05 MPa, 1.74 $\pm$ 1.15 MPa, and 1.81 $\pm$ 1.21 MPa under dynamic, intermediate, and quasi-static loading, respectively. Similarly, the elastic modulus for the circumferential sample was 3.15 $\pm$ 1.73 MPa, 2.14 $\pm$ 1.3 MPa, and 0.65 $\pm$ 1.25 MPa for the dynamic, intermediate, and quasi-static loading, respectively. Egorov et al. [24] conducted a similar investigation on the human colon using transverse specimens under quasi-static stress conditions. They recorded an ultimate stress value of 0.9 MPa, with a destructive strain of 180% (Fig. 6.5).

6.7 Vagina

The vaginal tissue is a smooth, elastic, and muscular tube that serves as an entrance to the female reproductive organs. This tissue also acts as a delivery channel for childbirth and pathway for the menstrual blood [27, 28]. The vaginal tissue is about 7.5 cm along the anterior wall and 9 cm along the posterior wall [29]. Goh [30] estimated the elastic properties of vaginal tissue and stated that postmenopausal women's

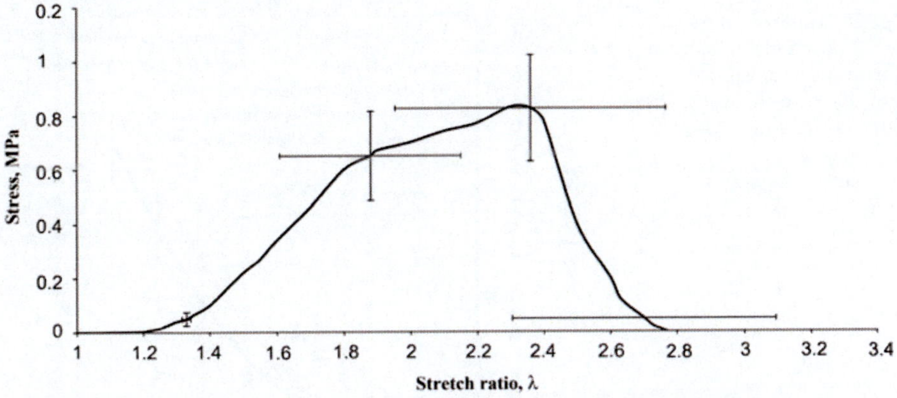

Fig. 6.5 Mechanical testing of human cadaveric large intestine transversal specimens [24]

vaginal tissue (14.35 MPa) had a slightly higher elastic modulus than premenopausal women's vaginal tissue (11.5 MPa). Significant work was reported by Lei et al. [31] comparing biomechanical properties of vaginal tissue for the prolapsed and non-prolapsed pre- and postmenopausal women. The elastic modulus, Poisson's ratio, and ultimate stress of premenopausal women with prolapse were recorded to be 9.45 ± 0.70 MPa, 0.43 ± 0.01, and 0.60 ± 0.02 MPa, respectively, whereas the corresponding values for premenopausal women without prolapse were 6.65 ± 1.48 MPa, 0.46 ± 0.01, and 0.79 ± 0.05 MPa. Similarly, postmenopausal women with prolapse had an elastic modulus of 12.10 ± 1.10 MPa, Poisson's ratio of 0.39 ± 0.01, and ultimate stress of 0.27 ± 0.03 MPa, whereas those without prolapse had values of 10.26 ± 1.10 MPa, 0.42 ± 0.01, and 0.42 ± 0.03 MPa, respectively.

Martins et al. [32] reported the longitudinal and transverse elastic moduli of vaginal tissue to be 6.2 ± 1.5 MPa and 5.4 ± 1.1 MPa, respectively. The respective ultimate stress value was 2.3 ± 0.5 MPa and 2.6 ± 0.9 MPa for the longitudinal and transverse axis. It was observed that the stretch ratios directly affect the elastic modulus of the vaginal tissue. For low values of stretch ratios, the modulus of elasticity ranges between 0.34 MPa and 7.91 MPa, but for high stretch ratios, it rises significantly within the range of 3.86 MPa to 58.88 MPa [33].

Baah-Dwomoh et al. [29] determined that the elastic modulus of vaginal tissue varies from 2.5 MPa to 30 MPa. Rubod et al. [35] performed a series of experiments on human vaginal tissue and quantified the stress range at rupture for prolapsed and non-prolapsed tissue. The estimated values were 2.12 MPa to 6.06 MPa for the prolapsed and 0.82 MPa to 2.62 MPa for the non-prolapsed vaginal tissue. The mechanical properties would be indispensable for the development of computational models with realistic mechanical properties for various applications, such as surgical planning for urogynecological surgeons and finite element analysis of pelvic organ prolapse. Figure 6.6 shows the uniaxial stress versus stretch plot of sutured prosthetic mesh, vaginal tissue, and prosthetic mesh at different strain rates [34].

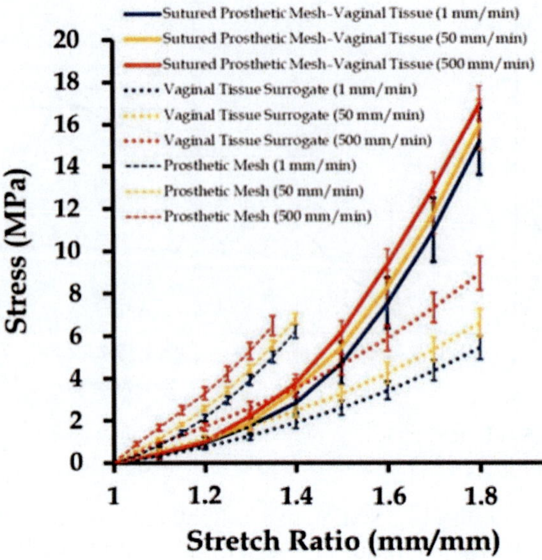

Fig. 6.6 Uniaxial stress versus stretch plot of sutured prosthetic mesh, vaginal tissue, and prosthetic mesh at different strain rates [34]

6.8 Urinary Bladder

The urinary bladder is a hollow organ that collects and excretes urine [36], located near the base of the pelvis and in the midline of the body [37]. Urine enters the urinary bladder via the ureters, gradually fill the urinary bladder and stretch its elastic walls. These walls enable the bladder to expand and hold around 600–800 ml of urine. The wall thickness of the urinary bladder varies from 3 mm $\pm$ 1 mm (female) to 3.3 mm $\pm$ 1.1 mm (males) [38].

To date, the studies on the investigation of mechanical properties of the bladder tissue are limited. A notable work was performed by Dahms et al. [39] who compared the mechanical properties of human, rat, and pig bladder tissue. The reported elastic modulus and ultimate tensile stresses of the human bladder were 0.25 MPa and 0.27 MPa, respectively. In addition, it was also stated that the ultimate tensile stress and elastic modulus of pig bladder were comparable to those of human bladder [39]. Martins et al. [40] evaluated the elastic modulus and maximum tensile strength of human female bladder tissue using uniaxial tensile testing. The 13 female cadaveric bladder samples were analyzed and categorized as younger females (below 50 years age) and older females (above 50 years age). It was found that the females from younger group possess more stiffness with an elastic modulus of 2.1 $\pm$ 0.2 MPa and ultimate stress of 1.0 $\pm$ 0.2 MPa, whereas the females from the older group showed an elastic modulus of 1.3 $\pm$ 0.1 MPa and ultimate stress of 0.7 $\pm$ 0.1 MPa. The urinary bladder is a complex tissue that connects with several other tissue organs [41], and to date, its mechanical properties are not well understood. Figure 6.7 illustrates the bladder tissue samples for the mechanical testing [36].

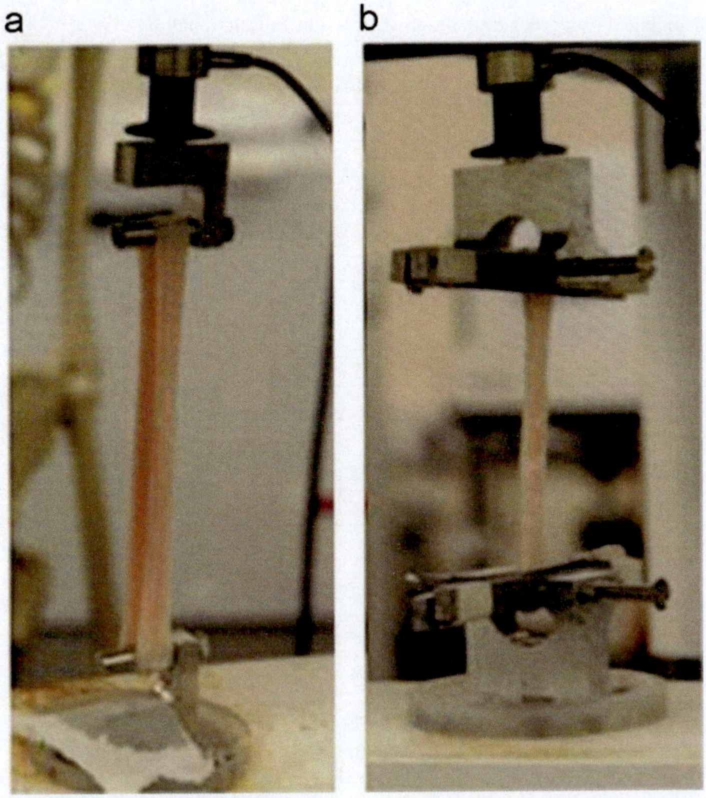

Fig. 6.7 Tissue samples for the testing of bladder tissue **a** looped samples; **b** rectangular samples [36]

6.9 Summarizing Mechanical Properties of Tissues with High Stiffness

Table 6.1 outlines the mechanical properties of human soft tissues in functional organs with low stiffness. The values of Table 6.1 were used to plot a graph, and Fig. 6.8 illustrates the modulus of elasticity (mean ± S.D.) of the discussed soft tissues.

Table 6.1 Mechanical properties of the high stiff tissues in functional organs

Tissue	References	Modulus of elasticity	Shear modulus	Ultimate tensile strength	Subgroup/remarks
Nasal cavity	[5]	1.39 MPa	–	–	Septal cartilage
	[7]	–	–	3.88 MPa	Nasal periosteum
	[6]	2.09 ± 0.81 MPa	–	–	Alar cartilages
	[6]	2.72 ± 0.63 MPa	–	–	Septal cartilage
Oral cavity	[8]	37.36 ± 17.4 MPa	–	3.81 ± 0.9 MPa	Gingiva
	[8]	18.13 ± 4.51 MPa	–	1.70 ± 0.9 MPa	Hard palate
	[8]	8.33 ± 5.78 MPa	–	1.54 ± 0.5 MPa	Buccal mucosa
	[9]	2.63 ± 1.08 kPa	–	–	Uvula
	[9]	6.54 ± 3.52 kPa	–	–	Soft palate
	[11]	0.9975 ± 0.48 kPa	–	–	Soft palate
	[12]	–	2.42 kPa	–	Soft palate
Heart	[13]	16.05 ± 2.02 MPa	–	2.78 ± 1.05 MPa	Pulmonary valve
	[13]	15.34 ± 3.84 MPa	–	1.74 ± 0.29 MPa	Aortic valve
	[14]	1.27 ± 0.63 to 27.9 ± 10.59 kPa	–	–	–
Pancreas	[19]	–	1.65 kPa	–	–
Small intestine	[25]	5.16 ± 3.03 MPa	–	–	Embalmed intestinal samples
	[25]	2.69 ± 0.37 MPa	–	–	Fresh intestinal samples
	[24]	–	–	0.9 MPa	–
Colon	[26]	3.16 ± 1.89 MPa	–	–	Dynamic loading
	[26]	1.94 ± 1.22 MPa	–	–	Intermediate loading
	[26]	1.19 ± 1.23 MPa	–	–	Quasi-static loading
	[24]	–	–	0.9 MPa	–
Vagina	[30]	14.35 MPa	–	–	Postmenopausal
	[30]	11.5 MPa	–	–	Premenopausal
	[32]	5.8 ± 1.3 MPa	–	2.4 ± 0.7 MPa	–
	[29]	2.5–30 MPa	–	–	–
Urinary bladder	[39]	0.25 MPa	0.27 MPa	–	–
	[40]	1.9 ± 0.2 MPa	0.9 ± 0.1 MPa	–	–

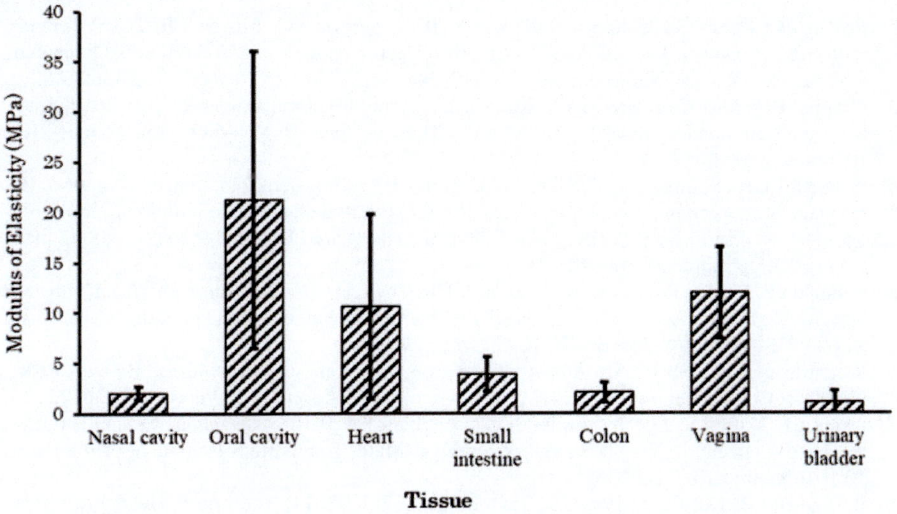

Fig. 6.8 Modulus of elasticity (mean ± S.D.) of the low stiff tissues in functional organs

References

1. Yu S, Sun XZ, Liu YX. Numerical analysis of the relationship between nasal structure and its function. Sci World J 2014;2014. https://doi.org/10.1155/2014/581975.
2. Rotter N, Tobias G, Lebl M, Roy AK, Hansen MC, Vacanti CA et al (2002) Age-related changes in the composition and mechanical properties of human nasal cartilage. Arch Biochem Biophys 403:132–140. https://doi.org/10.1016/S0003-9861(02)00263-1
3. Richmon JD, Sage A, Wong VW, Chen AC, Sah RL, Watson D (2006) Compressive Biomechanical Properties of Human Nasal Septal Cartilage. Am J Rhinol 20:496–501. https://doi.org/10.2500/ajr.2006.20.2932
4. Richmon JD, Sage AB, Wong VW, Chen AC, Pan C, Sah RL et al (2005) Tensile Biomechanical Properties of Human Nasal Septal Cartilage. Am J Rhinol 19:617–622. https://doi.org/10.1177/194589240501900616
5. Alkan Z, Yigit O, Acioglu E, Bekem A, Azizli E, Kocak I et al (2011) Tensile characteristics of costal and septal cartilages used as graft materials. Arch Facial Plast Surg 13:322–326. https://doi.org/10.1001/archfacial.2011.54
6. Griffin MF, Premakumar Y, Seifalian AM, Szarko M, Butler PEM (2016) Biomechanical characterisation of the human nasal cartilages; implications for tissue engineering. J Mater Sci Mater Med 27:1–6. https://doi.org/10.1007/s10856-015-5619-8
7. Zeng Y-J, Sun X, Yang J, Wu W, Xu X, Yan Y (2003) Mechanical properties of nasal fascia and periosteum. Clin Biomech 18:760–764. https://doi.org/10.1016/S0268-0033(03)00136-0
8. Choi JJE, Zwirner J, Ramani RS, Ma S, Hussaini HM, Waddell JN, et al. Mechanical properties of human oral mucosa tissues are site dependent: A combined biomechanical, histological and ultrastructural approach. Clin Exp Dent Res 2020:cre2.305. https://doi.org/10.1002/cre2.305.
9. Haddad SMH, Dhaliwal SS, Rotenberg BW, Samani A, Ladak HM (2018) Estimation of the Young's moduli of fresh human oropharyngeal soft tissues using indentation testing. J Mech Behav Biomed Mater 86:352–358. https://doi.org/10.1016/j.jmbbm.2018.07.004
10. Malhotra A, Huang Y, Fogel RB, Pillar G, Edwards JK, Kikinis R et al (2002) The male predisposition to pharyngeal collapse: Importance of airway length. Am J Respir Crit Care Med 166:1388–1395. https://doi.org/10.1164/rccm.2112072
11. Birch MJ, Srodon PD (2009) Biomechanical properties of the human soft palate. Cleft Palate-Craniofacial J 46:268–274. https://doi.org/10.1597/08-012.1

12. Brown EC, Cheng S, McKenzie DK, Butler JE, Gandevia SC, Bilston LE (2015) Tongue Stiffness is Lower in Patients with Obstructive Sleep Apnea during Wakefulness Compared with Matched Control Subjects. Sleep 38:537–544. https://doi.org/10.5665/sleep.4566
13. Stradins P (2004) Comparison of biomechanical and structural properties between human aortic and pulmonary valve*1. Eur J Cardio-Thoracic Surg 26:634–639. https://doi.org/10.1016/j.ejcts.2004.05.043
14. Holzapfel GA, Sommer G, Gasser CT, Regitnig P (2005) Determination of layer-specific mechanical properties of human coronary arteries with nonatherosclerotic intimal thickening and related constitutive modeling. Am J Physiol - Hear Circ Physiol 289:2048–2058. https://doi.org/10.1152/ajpheart.00934.2004
15. Canham PB, Finlay HM, Dixon JG, Boughner DR, Chen A (1989) Measurements from light and polarised light microscopy of human coronary arteries fixed at distending pressure. Cardiovasc Res 23:973–982. https://doi.org/10.1093/cvr/23.11.973
16. Richardson PD, Keeny SM. Anisotropy of human coronary artery intima. Bioeng. Proc. Northeast Conf., Publ by IEEE; 1989, p. 205–6. https://doi.org/10.1109/nebc.1989.36772.
17. Oveissi F, Naficy S, Lee A, Winlaw DS, Dehghani F (2020) Materials and manufacturing perspectives in engineering heart valves: a review. Mater Today Bio 5:100038. https://doi.org/10.1016/j.mtbio.2019.100038
18. Baillargeon B, Rebelo N, Fox DD, Taylor RL, Kuhl E (2014) The Living Heart Project: A robust and integrative simulator for human heart function. Eur J Mech - A/Solids 48:38–47. https://doi.org/10.1016/J.EUROMECHSOL.2014.04.001
19. Shi Y, Glaser KJ, Venkatesh SK, Ben-Abraham EI, Ehman RL (2015) Feasibility of using 3D MR elastography to determine pancreatic stiffness in healthy volunteers. J Magn Reson Imaging 41:369–375. https://doi.org/10.1002/jmri.24572
20. Wex C, Fröhlich M, Brandstädter K, Bruns C, Stoll A (2015) Experimental analysis of the mechanical behavior of the viscoelastic porcine pancreas and preliminary case study on the human pancreas. J Mech Behav Biomed Mater 41:199–207. https://doi.org/10.1016/j.jmbbm.2014.10.013
21. Bellini C, Glass P, Sitti M, Di Martino ES (2011) Biaxial mechanical modeling of the small intestine. J Mech Behav Biomed Mater 4:1727–1740. https://doi.org/10.1016/j.jmbbm.2011.05.030
22. Miyasaka EA, Okawada M, Utter B, Mustafa-Maria H, Luntz J, Brei D et al (2010) Application of distractive forces to the small intestine: Defining safe limits. J Surg Res 163:169–175. https://doi.org/10.1016/j.jss.2010.03.060
23. Gregersen H, Kassab G (1996) Biomechanics of the gastrointestinal tract. Neurogastroenterol Motil 8:277–297. https://doi.org/10.1111/j.1365-2982.1996.tb00267.x
24. Egorov VI, Schastlivtsev IV, Prut EV, Baranov AO, Turusov RA (2002) Mechanical properties of the human gastrointestinal tract. J Biomech 35:1417–1425. https://doi.org/10.1016/S0021-9290(02)00084-2
25. Bourgouin S, Bège T, Masson C, Arnoux PJ, Mancini J, Garcia S et al (2012) Biomechanical characterisation of fresh and cadaverous human small intestine: Applications for abdominal trauma. Med Biol Eng Comput 50:1279–1288. https://doi.org/10.1007/s11517-012-0964-y
26. Massalou D, Masson C, Afquir S, Baqué P, Arnoux PJ, Bège T (2019) Mechanical effects of load speed on the human colon. J Biomech 91:102–108. https://doi.org/10.1016/j.jbiomech.2019.05.012
27. Chanda A, Unnikrishnan V, Roy S, Richter HE. Computational Modeling of the Female Pelvic Support Structures and Organs to Understand the Mechanism of Pelvic Organ Prolapse: A Review. Appl Mech Rev 2015;67. https://doi.org/10.1115/1.4030967.
28. Ashton-Miller JA, Delancey JOL. Functional Anatomy of the Female Pelvic Floor. Ann N Y Acad Sci 2007;1101. https://doi.org/10.1196/annals.1389.034.
29. Baah-Dwomoh A, McGuire J, Tan T, De Vita R. Mechanical Properties of Female Reproductive Organs and Supporting Connective Tissues: A Review of the Current State of Knowledge. Appl Mech Rev 2016;68. https://doi.org/10.1115/1.4034442.

30. Goh JTW (2002) Biomechanical properties of prolapsed vaginal tissue in pre- and postmenopausal women. Int Urogynecol J 13:76–79. https://doi.org/10.1007/s001920200019
31. Lei L, Song Y, Chen RQ (2007) Biomechanical properties of prolapsed vaginal tissue in pre- and postmenopausal women. Int Urogynecol J 18:603–607. https://doi.org/10.1007/s00192-006-0214-7
32. Martins PALS, Jorge RMN, Ferreia AJM, Saleme CS, Roza T, Parente MMP et al (2011) Vaginal Tissue Properties versus Increased Intra-Abdominal Pressure: A Preliminary Biomechanical Study. Gynecol Obstet Invest 71:145–150. https://doi.org/10.1159/000315160
33. Chanda A, Flynn Z, Unnikrishnan V. Biomechanical Characterization of Normal and Prolapsed Vaginal Tissue Surrogates. J Mech Med Biol 2018;18. https://doi.org/10.1142/S0219519417501007.
34. Chanda A, Ruchti T, Upchurch W. Biomechanical Modeling of Prosthetic Mesh and Human Tissue Surrogate Interaction. Biomimetics 2018;3. https://doi.org/10.3390/biomimetics3030027.
35. Rubod C, Boukerrou M, Brieu M, Jean-Charles C, Dubois P, Cosson M (2008) Biomechanical properties of vaginal tissue: Preliminary results. Int Urogynecol J 19:811–816. https://doi.org/10.1007/s00192-007-0533-3
36. Barnes SC, Shepherd DET, Espino DM, Bryan RT (2015) Frequency dependent viscoelastic properties of porcine bladder. J Mech Behav Biomed Mater 42:168–176. https://doi.org/10.1016/j.jmbbm.2014.11.017
37. Chanda A, Unnikrishnan V (2018) Effect of bladder and rectal loads on the vaginal canal and levator ani in varying pelvic floor conditions. Mech Adv Mater Struct 25:1214–1223. https://doi.org/10.1080/15376494.2017.1331629
38. Ajalloueian F, Lemon G, Hilborn J, Chronakis IS, Fossum M. Bladder biomechanics and the use of scaffolds for regenerative medicine in the urinary bladder. Nat Rev Urol 2018;15. https://doi.org/10.1038/nrurol.2018.5.
39. Dahms SE, Piechota HJ, Dahiya R, Lue TF, Tanagho EA (1998) Composition and biomechanical properties of the bladder acellular matrix graft: Comparative analysis in rat, pig and human. Br J Urol 82:411–419. https://doi.org/10.1046/j.1464-410X.1998.00748.x
40. Martins PALS, Filho ALS, Fonseca AMRMI, Santos A, Santos L, Mascarenhas T, et al. Uniaxial mechanical behavior of the human female bladder. Int Urogynecol J 2011;22:991–5. https://doi.org/10.1007/s00192-011-1409-0.
41. Roccabianca S, Bush TR (2016) Understanding the mechanics of the bladder through experiments and theoretical models: Where we started and where we are heading. Technology 04:30–41. https://doi.org/10.1142/s2339547816400082

Chapter 7
Hyperelastic Models for Anisotropic Tissue Characterization

7.1 Introduction

Soft tissues are primarily heterogeneous and demonstrate regional and directional anisotropy in three dimensions, including the skin, skeletal muscles, connective tissues, and tissues composing organs (such as the brain and cardiac tissues). Soft tissues are primarily inhomogeneous and display spatial and directional anisotropy in three dimensions, including the epidermis, musculoskeletal system, connective tissues, and tissues constituting organs (such as the brain and cardiac tissues) [1–3]. The differences in the distribution of collagen fibers from one part of the body to another are primarily responsible for tissue anisotropy. In the past, histological analyzes and dissection were used to study collagen fiber dispersion in human bodies and animal models [4, 5]. Recent developments in imaging technology have made it possible to use confocal imaging and diffusion tensor magnetic resonance imaging (DT-MRI) to examine the distribution of fibers in the human body [6]. Very few attempts to simulate tissue anisotropy in finite element (FE) models of organs have been made in the literature [6–8]. For the assessment of tissue material in other FE and experimental models, isotropic and transversely-isotropic hyperelastic models have been used the majority of the time.

The strain-energy function (ψ), which is dependent on the kind of material, is used to describe the material behavior of an isotropic soft material. The strain energy, which is dependent on the kind of material, is used to describe the material behavior of an isotropic soft material (Fig. 7.1a) [9, 10]. The strain-energy function in an isotropic hyperelastic model depends on the principal stretches (λ_1, λ_2, and λ_3) or the Cauchy-green tensor invariants (I_1, I_2, and I_3) which are also functions of the principal stretches) [11] as shown in Eqs. 7.1 and 7.2. The well-known isotropic hyperelastic constitutive models such as Neo-Hookean, Mooney-Rivlin, Yeoh, Ogden, Humphrey, Martins, and Veronda-Westmann are used to describe soft tissues [10].

$$\psi_{\text{Isotropic}} = \psi(I_1, I_2, I_3) \tag{7.1}$$

A. Chanda and G. Singh, *Mechanical Properties of Human Tissues*,
Materials Horizons: From Nature to Nanomaterials,
https://doi.org/10.1007/978-981-99-2225-3_7

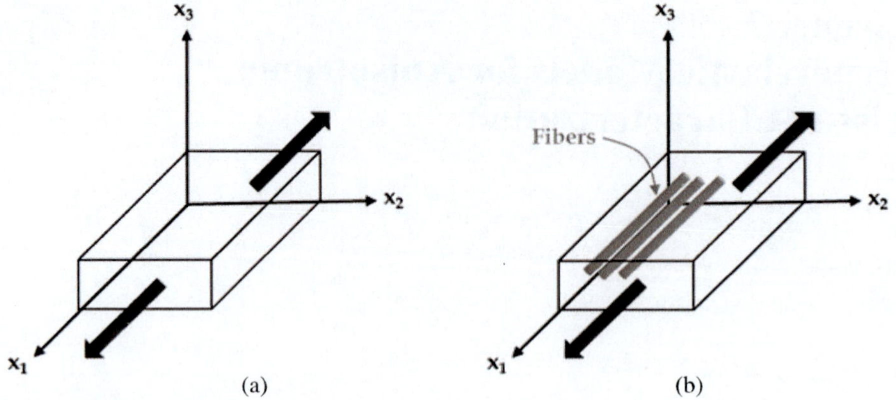

Fig. 7.1 Schematic of soft composite stretching in: **a** isotropic hyperelastic model and **b** transversely-isotropic hyperelastic model

$$I_1 = \sum_{i=1}^{3} \lambda_i^2,\ I_2 = \sum_{i,j=1}^{3} \lambda_i^2 \lambda_j^2,\ I_3 = \prod_{i=1}^{3} \lambda_i^2 \tag{7.2}$$

The first hyperelastic material model to be created, the Mooney-Rivlin model was effective in accurately fitting polynomial curves to mechanical stress–strain test data collected for nonlinear isotropic rubber-like materials. This model is shown in Eq. 7.3, where the imposed stress in a uniaxial test is expressed in terms of "λ" the uniaxial stretch, and "c_1" and "c_2" the two material constants that are predicted from curve fitting. In the past, the Mooney-Rivlin model has been widely used to forecast how soft tissue materials will behave in uniaxial and biaxial tests [12–14]. A reduced polynomial version of the Mooney-Rivlin model, the Yeoh model, which is depicted in Eq. 7.4, may precisely predict the nonlinear behavior of incompressible rubber-like materials. Recently, the Yeoh model has been utilized to forecast the mechanical properties of breast and brain tissues [15, 16].

$$\sigma_{\text{Mooney}} = 2(\lambda^2 - \frac{1}{\lambda})(c_1 + c_2\frac{1}{\lambda}) \tag{7.3}$$

$$\sigma_{\text{Yeoh}} = 2(\lambda^2 - \frac{1}{\lambda})(c_1 + 2c_2(I_1 - 3) + 3c_3(I_1 - 3)^2) \tag{7.4}$$

Neo-Hookean model (Eq. 7.5) was created in 1948 to investigate the mechanical characteristics of vulcanized rubber. It is a simple hyperelastic model that resembles Hooke's law. It has been found to be ineffective at predicting material behavior under biaxial loads and to be appropriate for strains under 20%. Neo-Hookean model is still used to forecast nonlinear soft tissue activity even though Mooney-Rivlin model has surpassed it in terms of prediction accuracy [17]. The Ogden model (Eq. 7.6) has been frequently used to define biological tissues because it more closely predicts the mechanical properties of incompressible, isotropic, and stress rate independent rubber-like materials than the Mooney-Rivlin model and is stable at large strains [18–21].

$$\sigma_{\text{Hookean}} = 2(\lambda^2 - \frac{1}{\lambda})c_1 \tag{7.5}$$

$$\sigma_{\text{Ogden}} = c_1(\lambda^{c_2} - 2^{-1+c_2}\lambda^{-c_2/2}) + c_3(\lambda^{c_4} - 2^{-1+c_4}\lambda^{-c_4/2}) + c_5(\lambda^{c_6} - 2^{-1+c_6}\lambda^{-c_6/2}) \tag{7.6}$$

In 1987, Humphrey and Yin created a composite-based material model for the soft tissues of the myocardium. The homogenous matrix and the many families of non-interacting fibers that make up soft tissue were both thought to be capable of significant deformation. The matrix and fiber contribution data from the multiaxial stress–strain experiments were used to create a pseudo-strain-energy function that relied on the restricted structural knowledge [22, 23]. Anisotropic, incompressible, and nonlinear behavior of soft tissues could be precisely predicted by Humphrey and Yin's formulation (Eq. 7.7), which is an exponential function with two material constants [24–27].

$$\sigma_{\text{Humphrey}} = 2(\lambda^2 - \frac{1}{\lambda})c_1c_2e^{c_2(I_1-3)} \tag{7.7}$$

The transversely-isotropic and incompressible hyperelastic model of Humphrey and Yin were recently modified to become Martin's hyperelastic model. The overall strain-energy expression was expanded to include a new word that represents the passive strain energy that was stored in the muscle fibers as a result of the uniaxial trials [28]. Martins et al. used Eq. 7.8, i.e., resulting stress–strain relationship to predict the nonlinear behavior of pelvic tissues since it could forecast the behavior of soft tissue materials [29–31] Similar to Humphrey and Yin's model, the Veronda-Westmann model [32] is a composite-based model in which the energy density function is dependent on the contribution of the matrix and fibers. This model, which was created based on experimental research on cat skin, has been used to forecast the behavior of silicon rubber phantoms, liver tissues, and breast tissues [33–35].

$$\sigma_{\text{Martins}} = 2(\lambda^2 - \frac{1}{\lambda})c_1c_2e^{c_1(I_1-3)} + 2\lambda(\lambda - 1)c_3c_4e^{c_3(\lambda-1)^2} \tag{7.8}$$

$$\sigma_{\text{Veronda-Westmann}} = 2(\lambda^2 - \frac{1}{\lambda})c_1c_2(e^{c_2(I_1-3)} - \frac{1}{2\lambda}) \tag{7.9}$$

The effect of fibers is not taken into account by isotropic hyperelastic models, which can only simulate the soft tissue's overall mechanical characteristics. The transversely-isotropic hyperelastic material models take into account fiber contribution and effect in some circumstances. However, the majority of these models do not take into account fiber-matrix interactions, fiber orientations, or layer distributions.

In this work, a unique soft composite-based hyperelastic material model was created that takes a soft tissue's organization of many layers of fiber families with various orientations as its starting point. This model takes into account the unique contributions of each fiber and matrix in each layer, the impact of different fiber orientations between levels, and the interactions between the fibers and the matrix.

7.2 Anisotropic Hyperelastic Model

7.2.1 Numerical Model

To take into account the anisotropic effect of fiber additions in a matrix (soft composite model), with all fibers originally arranged in the vector direction of vector a_0 (Fig. 7.2), the energy function takes the form $\psi = \psi(C, a_0)$, where C is the Cauchy-Green deformation tensor. To use the invariant approach for modeling transverse isotropy, two additional invariants related to a_0 are defined as $I_4 = a_0.C.a_0 = \lambda_F^2$, $I_5 = a_0.C^2.a_0$ where λ_F is the fiber stretch [36, 37] The strain-energy function is rewritten as $\psi(C, a_0) = \psi(I_1, I_2, I_3, I_4, I_5)$. Both the matrix and the fibers are taken to be incompressible materials in order to streamline calculations. Due to the difference in stiffness between the matrix and the fibers, neither the matrix nor the fibers will deform uniformly. We use deformation tensor C_m in matrix to define the invariants in strain-energy function of the matrix given by $\psi^M = C_{10}^m(I_1^m - 3)$ where $I_1^m = tr(C_m)$, and the material model is neo-Hookean type [36, 37]. C_{10}^m is related to shear modulus of the matrix material. The strain energy stored in the fibers is modeled as a function of I_4, which is based on the assumption that the strain energy stored in the fibers depends mainly on the fiber elongation, where $\lambda_F = \sqrt{I_4}$ is the actual fiber stretch. To account for nonlinear dependence of strain energy on fiber stretch, and also to account for the three-dimensional aspect of the fiber, the strain energy stored in fibers is written as $\psi^F = C_{10}^m f(I_4)(I_1^f - 3)$ where $I_1^f = tr(C_f)$, and the material model is neo-Hookean type, C_f is the deformation tensor of fibers, and $C_{10}^m f(I_4)$ is related to stretch-dependent shear modulus of the fiber [36, 37].

Considering initial fiber direction a_0 along x_1 direction of rectangular Cartesian coordinate system (Fig. 7.2), i.e., $a_0 = e_1 = \begin{bmatrix} 1 \; 0 \; 0 \end{bmatrix}^T$. Assuming second arbitrary perpendicular axis e_2 such that $e_1.e_2 = 0$. Third axis $e_3 = e_1 \otimes e_2 = 0$. With a deformation F, the deformed fiber direction Fa_0 is rotated back to original fiber direction $a_0 = e_1$. Let the rotation tensors for this be R_1 which is independent of choice of e_2, and R_2 which depends on the e_2 axis. Using e_2 as the rotation axis, R_1Fe_2 is rotated back to e_1 - e_2 plane. The new deformation gradient tensor F^* is written as $F^* = R_2R_1F$. As R_1 and R_2 are rigid body rotations, $(F^*)^TF^*$ is the same as F^TF. F^* is used as starting point of the following multiplicative decomposition of deformation gradient [36, 37].

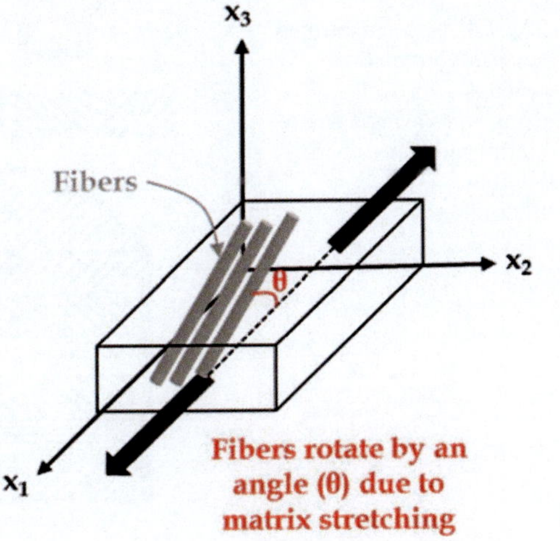

Fig. 7.2 Schematic of soft composite stretching in the novel soft composite-based anisotropic hyperelastic model formulation

From definition of F^* in the current deformation considered in Fig. 7.3, $F^*_{21} = F^*_{31} = F^*_{32} = 0$ and $F^*_{11} = |Fa_0| = \lambda_F$. F^* in e_1,e_2,e_3 coordinate system is given by Eq. 7.10 [36, 37].

$$F^* = \begin{bmatrix} \lambda_F & F^*_{12} & F^*_{13} \\ 0 & F^*_{22} & F^*_{23} \\ 0 & 0 & F^*_{33} \end{bmatrix} \tag{7.10}$$

The deformation gradient F^* is multiplicatively decomposed into uniaxial deformation gradient F^*_f in the fiber direction and subsequent shear deformation gradient F^*_s such that $F^* = F^*_s F^*_f$. F^*_f and F^*_s are given by Eq. 7.11 [36, 37].

$$\begin{aligned} F^*_f &= \mathrm{diag}\left[\lambda_F \; \lambda_F^{-\frac{1}{2}} \; \lambda_F^{-\frac{1}{2}} \right], \\ F^*_s &= F^*(F^*_f)^{-1} = \begin{bmatrix} 1 & F^*_{12}\lambda_F^{\frac{1}{2}} & F^*_{13}\lambda_F^{\frac{1}{2}} \\ 0 & F^*_{22}\lambda_F^{\frac{1}{2}} & F^*_{23}\lambda_F^{\frac{1}{2}} \\ 0 & 0 & F^*_{33}\lambda_F^{\frac{1}{2}} \end{bmatrix} \end{aligned} \tag{7.11}$$

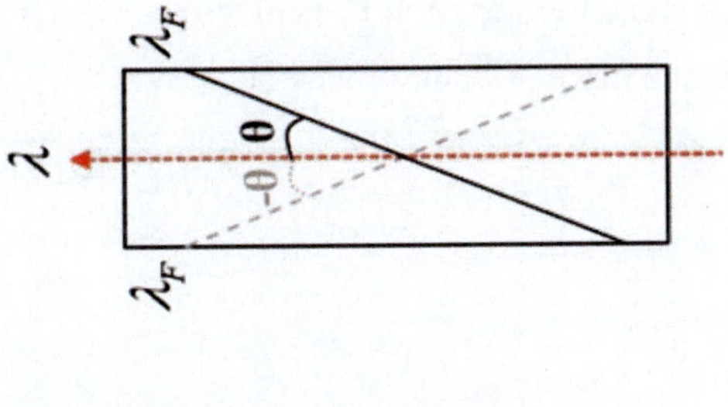

Fig. 7.3 Diagram illustrating the stretching phenomenon along the fiber (λ_F) which is at an angle $+\theta$ or $-\theta$ with respect to the soft composite uniaxial stretch λ

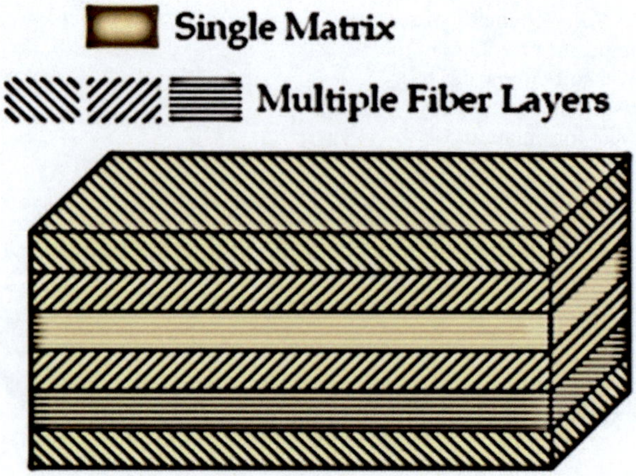

Fig. 7.4 New anisotropic material formulation schematic for soft composites using a single matrix and many fiber families

By suitable choice of e_2 axis, $F^*_{22}\lambda_F^{\frac{1}{2}} = 1$. F^*_{12} and F^*_{13} are along fiber shears, and F^*_{23} is termed as transverse shear. Shear deformations are related to invariants of C given by Eq. 7.12 [36, 37].

$$\left(F^*_{12}\right)^2 + \left(F^*_{13}\right)^2 = \frac{I_5 - I_4^2}{I_4},\ \left(F^*_{23}\right)^2 = I_1 - \frac{I_5 + 2\sqrt{I_4}}{I_4} \tag{7.12}$$

7.2.2 Modeling the Effect of Fiber and Matrix Contributions

Soft tissues have been thoroughly researched in the literature, and it has been discovered that they display a wide variety of stress–strain behavior because collagen fiber concentration and distribution can vary. Connective tissues have been found to contain the highest amounts of collagen fiber, making them much stronger than other tissues. Additionally, alterations in soft (pelvic) tissue have been linked to differences in collagen fiber due to conditions like pelvic organ prolapse (POP). To model the effect of individual fiber and matrix contribution on a soft composite material property, the matrix volume fraction (MVF) v_m and fiber volume fraction (FVF) v_f terms were introduced. The strain energy stored in the whole composite during uniaxial deformation F^*_f is given by Eq. 7.13, where $I_1(F^*_f) = \lambda_F^2 + \frac{2}{\lambda_F} = I_4 + 2I_4^{-\frac{1}{2}}$ is the first invariant of $C^*_f = \left(F^*_f\right)^T F^*_f$.

$$\psi_f = \psi_f^M + \psi_f^F = v_m C_{10}^m (I_1(F^*_f) - 3) + v_f C_{10}^m f(I_4)(I_1(F^*_f) - 3) \tag{7.13}$$

On application of shear deformation F_s^*, fiber stretch and stiffness remain constant. In this step, the composite is treated as neo-Hookean matrix embedded with neo-Hookean fibers where $f(I_4)$ is ratio of the fiber and matrix stiffnesses. The quantity of shear deformation is computed by adding the terms as shown in Eq. 7.14, where $I_1(F_f^*) = I_4 + 2I_4^{-\frac{1}{2}}$.

$$\left(F_{12}^*\right)^2 + \left(F_{13}^*\right)^2 + \left(F_{23}^*\right)^2 = \frac{I_5 - I_4}{I_4} + I_1 - \frac{I_5 + 2\sqrt{I_4}}{I_4} = I_1 - I_1(F_f^*) \quad (7.14)$$

The energy stored during deformation F_s^* of the neo-Hookean composite is given by Eq. 7.15, where c_3 is the effective neo-Hookean modulus, and $I_1 - I_1(F_f^*)$ is the portion of I_1 remaining after deformation F_s^* [36, 37].

$$\psi_s = c_3\left[I_1 - I_1(F_f^*)\right] \quad (7.15)$$

Now, strain-energy function for the anisotropic composite model is sum of the strain energies of the two steps (Eqs. 7.13 and 7.15), which is given by Eq. 7.16 [36, 37]. In Eq. 7.16, C_{10}^m and $C_{10}^m f(I_4)$ are substituted with constants c_1 and c_2, respectively.

$$\begin{aligned}\psi =&\psi_f + \psi_s = v_f c_2\left[I_1(F_f^*) - 3\right] + v_m c_1\left[I_1(F_f^*) - 3\right] \\ &+ c_3\left[I_1 - I_1(F_f^*)\right]\end{aligned} \quad (7.16)$$

For Humphrey model fibers and matrix definitions, the modified strain-energy functions and the net strain-energy functions are given by Eqs. 7.17 and 7.18, respectively.

$$\psi^M = c_1\left[e^{c_2(I_1^m - 3)} - 1\right], \ \psi^F = c_3\left[e^{c_4(I_1^f - 3)} - 1\right] \quad (7.17)$$

$$\begin{aligned}\psi =&v_m c_1\left[e^{c_2(I_1(F_f^*) - 3)} - 1\right] + v_f c_3\left[e^{c_4(I_1(F_f^*) - 3)} - 1\right] \\ &+ c_5\left[I_1 - I_1(F_f^*)\right]\end{aligned} \quad (7.18)$$

In Eqs. 7.16 and 7.18, the term $\left[I_1 - I_1(F_f^*)\right]$ goes to zero with $I_1(F_f^*) = I_1$ as λ_F (the stretch in the direction of the fiber) in case of a uniaxial tensile test coincides with the applied stretch λ for a fiber direction along the direction of stretch. Also, substituting $I_1(F_f^*) = \lambda^2 + \frac{2}{\lambda}$, Eqs. 7.16 and 7.18 can be rewritten as Eqs. 7.19 and 7.20, respectively.

$$\psi_{\text{Neo-Hookean/Modified}} = (v_m c_1 + v_f c_2)(\lambda^2 + \frac{2}{\lambda} - 3) \quad (7.19)$$

$$\psi_{\text{Humphrey/Modified}} = v_{\text{m}} c_1 \left[e^{c_2(\lambda^2 + \frac{2}{\lambda} - 3)} - 1 \right] + v_{\text{f}} c_3 \left[e^{c_4(\lambda^2 + \frac{2}{\lambda} - 3)} - 1 \right] \tag{7.20}$$

The principal Cauchy stress for a given hyperelastic strain-energy function was found using Eq. 7.21, where $\sigma_2 = \sigma_3 = 0$ are the boundary conditions for the uniaxial tension experiment. Here, $\lambda_1 = \lambda$ and $\lambda_3 = 0$.

$$\sigma_1 = \lambda_1 \frac{\partial \psi}{\partial \lambda_1} - \lambda_3 \frac{\partial \psi}{\partial \lambda_3}, \sigma_2 = \sigma_3 = 0 \tag{7.21}$$

Substituting the strain-energy functions from Eqs. 7.19 and 7.20 into Eq. 7.21, the stress versus stretch equations are given by Eqs. 7.22 and 7.23.

$$\sigma_{\text{Neo-Hookean/Modified}} = 2(v_{\text{m}} c_1 + v_{\text{f}} c_2)(\lambda - \frac{1}{\lambda^2}) \tag{7.22}$$

$$\sigma_{\text{Humphrey/Modified}} = 2 v_{\text{m}} c_1 c_2 (\lambda - \frac{1}{\lambda^2}) \left[e^{c_2(\lambda^2 + \frac{2}{\lambda} - 3)} \right] + 2 v_{\text{f}} c_3 c_4 (\lambda - \frac{1}{\lambda^2}) \left[e^{c_4(\lambda^2 + \frac{2}{\lambda} - 3)} \right] \tag{7.23}$$

Material characterization is conducted using Eqs. 22 and 23 by substituting the fiber contribution (v_{f}) and matrix contribution (v_{m}) in the stress-stretch equations and estimating the values of the constants.

7.2.3 Modeling the Effect of Fiber Orientation

To account for single layer fiber orientation effect in the soft composites (Fig. 7.3), $\lambda_{\text{F}} = \lambda \cos\theta$ substitution is conducted in Eqs. 22 and 23, so that $I_1(\text{F}_{\text{f}}^*) = \lambda_{\text{F}}^2 + \frac{2}{\lambda_{\text{F}}} = \lambda^2 \cos^2\theta + \frac{2}{\lambda \cos\theta}$, $\left[I_1 - I_1(\text{F}_{\text{f}}^*)\right] = \left[\lambda^2(\sin^2\theta) + \frac{2}{\lambda}(1 - \sec\theta)\right]$, and $0 \leq \theta < 90°$. The modified stress–strain relationships corresponding to Eqs. 22 and 23 are Eqs. 7.24 and 7.25, respectively.

$$\sigma_{\text{Neo-Hookean/Modified2}} = 2(v_{\text{m}} c_1 + v_f c_2)(\lambda \cos\theta - \frac{1}{\lambda^2 \cos^2\theta}) + 2c_3 \left[\lambda(\sin^2\theta) - \frac{1}{\lambda^2}(1 - \sec\theta) \right] \tag{7.24}$$

$$\sigma_{\text{Humphrey/Modified2}} = 2v_{\text{m}}c_1c_2\left(\lambda\cos\theta - \frac{1}{\lambda^2\cos^2\theta}\right)\left[e^{c_2\left(\lambda^2\cos^2\theta + \frac{2}{\lambda\cos\theta} - 3\right)}\right] + 2v_{\text{f}}c_3c_4\left(\lambda\cos\theta - \frac{1}{\lambda^2\cos^2\theta}\right)\left[e^{c_4\left(\lambda^2\cos^2\theta + \frac{2}{\lambda\cos\theta} - 3\right)}\right] + 2c_5\left[\lambda(\sin^2\theta) - \frac{1}{\lambda^2}(1 - \sec\theta)\right] \quad (7.25)$$

7.2.4 Modeling the Effect of Multiple Fiber Layers at Arbitrary Orientations

The mechanical property of a soft composite with a single matrix and several fiber families implanted in it is solely reliant on the cumulative impact of fiber directions. For a composite with n fiber families where the fibers are oriented at angles $\theta_1, \theta_2, \theta_3, \ldots\theta_i\ldots\theta_n$ (where $0 \le \theta_i < 90°$ and $n > 1$), volume fractions are v_m for the matrix and $v_{\text{f1}}, v_{\text{f2}}, v_{\text{f3}}, \ldots v_{\text{f}i}\ldots v_{\text{f}n}$ for each fiber family, respectively, the homogenized strain-energy function can be written as Eq. 7.26.

$$\psi_{\text{Soft Composite}} = \sum_{i=1}^{n} \psi^i \quad (7.26)$$

The modified generic stress–strain relationships based on Eqs. 7.24, 7.25, and 7.26 for a soft tissue composite with multiple fiber families are given by Eqs. 7.27 and 7.28.

$$\sigma_{\text{Soft Composite: Neo-Hookean}} = \sum \left\{ \begin{array}{l} 2((1 - v_{\text{f}i})c_1 + v_{\text{f}i}c_2)\left(\lambda\cos\theta_i - \frac{1}{\lambda^2\cos^2\theta_i}\right) \\ +2c_3\left[\lambda(\sin^2\theta_i) - \frac{1}{\lambda^2}(1 - \sec\theta_i)\right] \end{array} \right\}, 1 \le i \quad (7.27)$$

$$\sigma_{\text{Soft Composite: Humphrey}} = \sum \left\{ \begin{array}{l} 2(1 - v_{fi})c_1c_2\left(\lambda\cos\theta_i - \frac{1}{\lambda^2\cos^2\theta_i}\right)\left[e^{c_2\left(\lambda^2\cos^2\theta_i + \frac{2}{\lambda\cos\theta_i} - 3\right)}\right] \\ +2v_{fi}c_3c_4\left(\lambda\cos\theta_i - \frac{1}{\lambda^2\cos^2\theta_i}\right)\left[e^{c_4\left(\lambda^2\cos^2\theta_i + \frac{2}{\lambda\cos\theta_i} - 3\right)}\right] \\ +2c_5\left[\lambda(\sin^2\theta_i) - \frac{1}{\lambda^2}(1 - \sec\theta_i)\right] \end{array} \right\}, \quad 1 \le i \le n \quad (7.28)$$

References

1. Picinbono G, Delingette H, Ayache N (2001) Nonlinear and anisotropic elastic soft tissue models for medical simulation. Proc 2001 ICRA. IEEE Int Conf Robot Autom (Cat. No. 01CH37164), IEEE. https://doi.org/10.1109/ROBOT.2001.932801
2. Chanda A, Unnikrishnan V, Roy S, Richter HE (2015) Computational modeling of the female pelvic support structures and organs to understand the mechanism of pelvic organ prolapse: a review. Appl Mech Rev 67. https://doi.org/10.1115/1.4030967/370016
3. Chanda A, Ghoneim H (2015) Pumping potential of a two-layer left-ventricle-like flexible-matrix-composite structure. Compos Struct 122:570–575. https://doi.org/10.1016/J.COMPSTRUCT.2014.11.069
4. Lowry OH, Gilligan DR, Katersky EM (1941) The determination of collagen and elastin in tissues, with results obtained in various normal tissues from different species. J Biol Chem 139:795–804. https://doi.org/10.1016/s0021-9258(18)72951-7
5. Neuman RE, Logan MA (1950) The determination of collagen and elastin in tissues. J Biol Chem 186:549–556. https://doi.org/10.1016/s0021-9258(18)56248-7
6. Basser PJ, Pajevic S, Pierpaoli C, Duda J, Aldroubi A (2000) In vivo fiber tractography using DT-MRI data. Magn Reson Med 44:625–632. https://doi.org/10.1002/1522-2594(200010)44:4
7. Wang Y, Haynor DR, Kim Y (2001) An investigation of the importance of myocardial anisotropy in finite-element modeling of the heart: methodology and application to the estimation of defibrillation efficacy. IEEE Trans Biomed Eng 48:1377–1389. https://doi.org/10.1109/10.966597
8. Colli Franzone P, Guerri L, Pennacchio M, Taccardi B (1998) Spread of excitation in 3-D models of the anisotropic cardiac tissue. II. Effects of fiber architecture and ventricular geometry. Math Biosci 147:131–71. https://doi.org/10.1016/S0025-5564(97)00093-X
9. Chanda A, Graeter R, Unnikrishnan V (2015) Effect of blasts on subject-specific computational models of skin and bone sections at various locations on the human body. AIMS Mater Sci 2:425–447. https://doi.org/10.3934/matersci.2015.4.425
10. Martins PALS, Jorge RMN, Ferreira AJM (2006) A comparative study of several material models for prediction of hyperelastic properties: Application to silicone-rubber and soft tissues. Strain 42:135–147. https://doi.org/10.1111/j.1475-1305.2006.00257.x
11. Beatty MF (1987) Topics in finite elasticity: hyperelasticity of rubber, elastomers, and biological tissues—with examples. Appl Mech Rev 40:1699–1734. https://doi.org/10.1115/1.3149545
12. Chabanas M, Payan Y, Marécaux C, Swider P, Boutault F (2004) Comparison of linear and non-linear soft tissue models with post-operative CT scan in maxillofacial surgery. Lect Notes Comput Sci (Including Subser Lect Notes Artif Intell Lect Notes Bioinformatics) 3078:19–27. https://doi.org/10.1007/978-3-540-25968-8_3/COVER
13. Guerin HL, Elliott DM (2007) Quantifying the contributions of structure to annulus fibrosus mechanical function using a nonlinear, anisotropic, hyperelastic model. J Orthop Res 25:508–516. https://doi.org/10.1002/JOR.20324
14. Hirokawa S, Tsuruno R (2000) Three-dimensional deformation and stress distribution in an analytical/computational model of the anterior cruciate ligament. J Biomech 33:1069–1077. https://doi.org/10.1016/S0021-9290(00)00073-7
15. Kaster T, Sack I, Samani A (2011) Measurement of the hyperelastic properties of ex vivo brain tissue slices. J Biomech 44:1158–1163. https://doi.org/10.1016/j.jbiomech.2011.01.019
16. O'Hagan JJ, Samani A (2009) Measurement of the hyperelastic properties of 44 pathological ex vivo breast tissue samples. Phys Med Biol 54:2557–2569. https://doi.org/10.1088/0031-9155/54/8/020
17. Miller K (2005) Method of testing very soft biological tissues in compression. J Biomech 38:153–158. https://doi.org/10.1016/J.JBIOMECH.2004.03.004
18. Velardi F, Fraternali F, Angelillo M (2006) Anisotropic constitutive equations and experimental tensile behavior of brain tissue. Biomech Model Mechanobiol 5:53–61. https://doi.org/10.1007/S10237-005-0007-9/METRICS

19. El Sayed T, Mota A, Fraternali F, Ortiz M (2008) A variational constitutive model for soft biological tissues. J Biomech 41:1458–1466. https://doi.org/10.1016/J.JBIOMECH.2008.02.023
20. Gao Z, Lister K, Desai JP (2010) Constitutive modeling of liver tissue: experiment and theory. Ann Biomed Eng 38:505–516. https://doi.org/10.1007/s10439-009-9812-0
21. Holzapfel GA, Gasser TC, Ogden RW (2000) A new constitutive framework for arterial wall mechanics and a comparative study of material models. J Elast 61:1–48. https://doi.org/10.1023/A:1010835316564/METRICS
22. Humphrey JD, Yin FCP (1987) On constitutive relations and finite deformations of passive cardiac tissue: I a pseudostrain-energy foundation. J Biomech Eng 109:298–304. https://doi.org/10.1115/1.3138684
23. Humphrey JD, Yin FC (1987) A new constitutive formulation for characterizing the mechanical behavior of soft tissues. Biophys J 52:563–570. https://doi.org/10.1016/S0006-3495(87)83245-9
24. Seshaiyer P, Humphrey JD (2003) a sub-domain inverse finite element characterization of hyperelastic membranes including soft tissues. J Biomech Eng 125:363–371. https://doi.org/10.1115/1.1574333
25. Van Loocke M, Lyons CG, Simms CK (2006) A validated model of passive muscle in compression. J Biomech 39:2999–3009. https://doi.org/10.1016/j.jbiomech.2005.10.016
26. Wang DHJ, Makaroun M, Webster MW, Vorp DA (2001) Mechanical properties and microstructure of intraluminal thrombus from abdominal aortic aneurysm. J Biomech Eng 123:536–539. https://doi.org/10.1115/1.1411971
27. Weiss JA, Maker BN, Govindjee S (1996) Finite element implementation of incompressible, transversely isotropic hyperelasticity. Comput Methods Appl Mech Eng 135:107–128. https://doi.org/10.1016/0045-7825(96)01035-3
28. Martins P, Peña E, Calvo B, Doblaré M, Mascarenhas T, Jorge RN et al (2010) Prediction of nonlinear elastic behaviour of vaginal tissue: experimental results and model formulation, vol 13, pp 327–37. https://doi.org/10.1080/10255840903208197
29. Martins PALS, Filho ALS, Fonseca AMRMI, Santos A, Santos L, Mascarenhas T et al (2011) Uniaxial mechanical behavior of the human female bladder. Int Urogynecol J 22:991–5. https://doi.org/10.1007/S00192-011-1409-0/TABLES/1
30. Peña E, Calvo B, Martínez MA, Martins P, Mascarenhas T, Jorge RMN et al (2010) Experimental study and constitutive modeling of the viscoelastic mechanical properties of the human prolapsed vaginal tissue. Biomech Model Mechanobiol 9:35–44. https://doi.org/10.1007/S10237-009-0157-2/METRICS
31. Peña E, Martins P, Mascarenhas T, Natal Jorge RM, Ferreira A, Doblaré M et al (2011) Mechanical characterization of the softening behavior of human vaginal tissue. J Mech Behav Biomed Mater 4:275–283. https://doi.org/10.1016/J.JMBBM.2010.10.006
32. Veronda DR, Westmann RA (1970) Mechanical characterization of skin—finite deformations. J Biomech 3:111–124. https://doi.org/10.1016/0021-9290(70)90055-2
33. Groves RB, Coulman SA, Birchall JC, Evans SL (2013) An anisotropic, hyperelastic model for skin: experimental measurements, finite element modelling and identification of parameters for human and murine skin. J Mech Behav Biomed Mater 18:167–180. https://doi.org/10.1016/j.jmbbm.2012.10.021
34. Mehrabian H, Campbell G, Samani A (2010) A constrained reconstruction technique of hyperelasticity parameters for breast cancer assessment. Phys Med Biol 55. https://doi.org/10.1088/0031-9155/55/24/007
35. Pavan TZ, Madsen EL, Frank GR, Adilton O Carneiro A, Hall TJ (2010) Nonlinear elastic behavior of phantom materials for elastography. Phys Med Biol 55:2679. https://doi.org/10.1088/0031-9155/55/9/017
36. Holzapfel GA, Ogden RW, Holzapfel GA, Ogden RW (2010) Constitutive modelling of arteries. Proc R Soc A Math Phys Eng Sci 466:1551–1597. https://doi.org/10.1098/RSPA.2010.0058
37. Brown LW, Smith LM (2011) A simple transversely isotropic hyperelastic constitutive model suitable for finite element analysis of fiber reinforced elastomers. J Eng Mater Technol 133. https://doi.org/10.1115/1.4003517/475345

Chapter 8
Applications, Challenges, and Future Opportunities

8.1 Applications

The human body is exposed to impact loading in a variety of circumstances, such as car accidents, falls, gunshots, and blast effects. Due to sudden mechanical impacts, trauma injuries temporarily or permanently damage the tissue. Such impacts may be due to falls, automotive crashes, etc., and affect the tissue functionality biologically as well as structurally. For example, in the last few years, the traumatic brain injuries are prone in the soldiers due to blast exposures. The evaluation of such studies can only be possible computationally and not experimentally. Trauma injuries are also a major concern in contact sports, where unexpected impacts are frequent [1–3]. Also, the mechanical properties are anticipated for the accurate mitigation of injuries. For example, millions of women around the globe have been affected by inadequately assessed gynecological surgeries and urogynecology mesh implantations.

In the recent years, computational modeling of biological systems is pioneering to investigate the different boundary conditions with realistic mechanical properties of the soft tissues. Different material models could be developed and tested computationally to estimate an appropriate solution for the concern problem statement. The experimental trials with tissue simulants can be time consuming and could be expensive to fabricate the model at a full-scale. For such cases, the computational modeling is helpful to design and evaluate the responses of complex biological systems without performing the actual experimentation. The biomechanical behavior provides stress–strain profiles, and the results can be checked by varying the boundary conditions of the designed system (Fig. 8.1).

Due to the unavailability of cadaveric tissues, the tissue simulants with realistic mechanical properties play an important role. A tissue surrogate with realistic biomechanical properties would be indispensable to imitate the surgical suturing practice, diagnosis purposes, making medical models for educational purposes. The tissue simulants are low-cost and easy to fabricate and carry without any ethical or biosafety. The human tissue simulants are also necessary in the sports industry to

A. Chanda and G. Singh, *Mechanical Properties of Human Tissues*,
Materials Horizons: From Nature to Nanomaterials,
https://doi.org/10.1007/978-981-99-2225-3_8

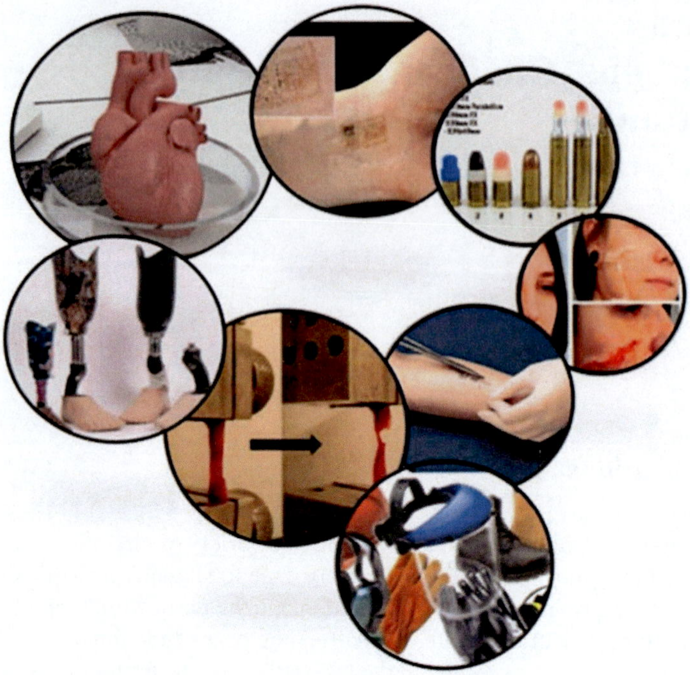

Fig. 8.1 Applications of soft tissue simulations

investigate a physical interface for affixing PPE and assessing actual damage risk. The tissue simulant exhibits comparable ultimate stress and elastic modulus similar to the human tissue and shows structural biofidelity. Such precisely characterized tissue simulants would be clinically helpful for surgical training, trauma research, understanding injury biomechanics, and developing the models for various diseases (e.g., liver cirrhosis, fatty liver, gallbladder biomechanics, Alzheimer, etc.).

8.2 Challenges

Across all the human soft tissues, only few soft tissues such as skin, muscles, and connective tissues are extensively studied human soft tissues. On the other hand, brain, heart, liver, lungs, excretory system organs, and reproductive system organs are the prominent functional tissues investigated to quantify their mechanical properties. To date, the mechanical properties of most of the abdominal tissues such as that of the digestive system and excretory systems have not studied extensively. The mechanical characteristics of the tongue, tonsils, salivary glands, esophagus, nasal cavity, and oral cavity have only been partially defined in research. In addition, there is no available data or material properties of the lymph nodes, glands (such as thyroid, adrenal), appendix, bone marrow, and a few crucial reproductive organs include the testis, ovary, epididymis, and fallopian tubes. The majority of studies on the mechanical properties of soft tissues have been linear and isotropic. To properly understand the behaviour of a soft tissue, hyperelastic curve fit models must be used to study and describe nonlinear and anisotropic features, directional metrics (such as stress against strain or force vs displacement) under various strain rates. To date, the variation of strain rates has not been extensively studied by the researchers, and limited strain rates were used to mechanically test the soft tissues. So far, isotropic and transversely-isotropic models have been used to evaluate the physical behavior of the skin, muscles, brain, breast, liver, spleen, pancreas, small intestine, and esophagus. There is a gap in the literature to study the soft tissue anisotropy of the functional tissues, effect of varying strain rates, and cyclic and dynamic loading conditions.

Currently, the mechanical testing of the soft tissues is a challenging task due to the limitations of tissue availability and the ethical issues related to the handling of cadaveric tissues. First, biosafety and ethical approvals are needed to perform the experimental testing and dispose-off the tissue samples (cadaveric or animal). Second, the mechanical properties of the cadaveric tissue samples dehydrate with time and significantly affect the testing results. Third, it is challenging to compare the findings because the majority of the existing research has focused on the mechanical characteristics of soft tissues under different strain rates. Fourth, there are differences between the data measured using in vivo and in vitro methods (e.g., while in vivo indentation experiments produce bulk elasticity measurements, in vitro tests yield dynamic stress versus strain responses). Fifth, the available imaging-based techniques and the in vivo methods (e.g., indentation) are limited to study a wide range of soft tissues. Sixth, most of the cadaveric soft tissue properties have been reported decades back when the test instruments had different testing methods and standards.

8.3 Future Opportunities

In future studies, standard testing protocols such as sample preparation, hydration content in tissue samples, constant strain rate, and measurement techniques may be helpful for evaluating the mechanical properties of the soft tissues. In addition, more sophisticated mechanical and imaging methods (e.g., DT-MRI) would be beneficial for precise and realistic characterization of the mechanical properties of the soft tissues. The knowledge of mechanical properties of the soft tissues would facilitate the development of artificial tissues, and these biofidelic tissues anticipated for the surgical training, ballistic testing, estimating the failure risk during the traumatic injuries. Future assessments of soft tissue mechanical characteristics may benefit from standard, uniform protocols that account for hydration and similar chemical treatments, a constant strain rate, and in vitro measurements for soft tissue preparation and testing. Soft tissues that have received little to no mechanical testing during the past ten years must be tested using the most recent protocols and techniques. Research on soft tissues using anisotropic hyperelastic models will enable precise computational modelling of such things to detect and avoid disorders, and the findings will serve as a crucial guide for loading and failure studies. Additionally, understanding the mechanical characteristics of the entire body's soft tissues can help with the creation of synthetic tissue substitutes and biofidelic simulators for surgical training.

8.3.1 Measurement of Mechanical Properties of Internal Organs in Normal Condition

Identification of the material characteristics that control the behaviour of soft biological tissues is known as soft tissue characterisation [4]. Researchers can more effectively use a material's unique characteristics by understanding its natural responses with the use of knowledge of these material attributes. Mechanical qualities, which are frequently quantified in engineering materials, are the type of soft tissue material properties that have been studied the most. Young's modulus, shear modulus, complex modulus, Poisson's ratio, yield strength, and ultimate strength are among the mechanical properties which are studied. Since mechanical properties are required to completely comprehend the mechanics of soft tissues, the characterization of soft tissues has many essential applications in the medical field. In order to provide a virtual reality-based environment where doctors may correctly educate and practise surgical methods, surgical simulators must also have accurate mechanical qualities [5].

Using characterization techniques to precisely assess the mechanical characteristics of soft tissues is fraught with challenges. The major problems with these techniques are the soft tissue's tiny size in comparison with other materials and their extreme fragility. They exhibit more minute mechanical characteristics, such as elastic characteristics that have been recorded in the tens of Pa to tens of kPa range. Another frequent problem with soft tissue characterization is sample preparation. Because they are very compliant materials with a surface profile that mirrors the underlying architecture of their component materials, soft tissues are also difficult to cut to precise dimensions. Historically, a wide range of various methods and characterisation techniques have been used to test the mechanical characteristics of soft tissues [6]. Even though these techniques can sufficiently characterize soft tissues, there are still many variations between the outcomes they provide.

8.3.2 Measurement of Mechanical Properties of Diseased and Damaged Tissues

Diseased and damaged tissues are a cause of serious concern for individuals. Understanding the biomechanical properties of diseased and damaged tissues will help in understanding the progression of such injured tissues and help in the development of remedial methods to heal such injured tissues. Recent focus in the category of diseases and damaged tissues is on brain tissues and articular cartilage. Understanding the cause of traumatic brain injury requires knowledge of the mechanical characteristics of brain tissue (TBI). Most studies during the past few decades have concentrated on the healthy brain tissues, whereas very few have examined the wounded regions. As a result, little is understood about the mechanical characteristics of the damaged brain tissues. For rest of the soft tissues, there is research gap, and further studies can be carried out to measure the mechanical properties of disease and damaged tissues.

8.3.3 Characterizing Tissue Anisotropy

Skin, skeletal muscles, connective tissues, and organ-forming tissues (such the brain and cardiac tissues) are examples of soft tissues in the human body that are neither homogeneous or isotropic [7, 8]. In three dimensions, these tissues display regional and directional anisotropy [9]. The differences in collagen fiber distribution in tissues may be primarily responsible for this material anisotropy. In the past, histological studies were used to examine the collagen fiber dispersion in human bodies and animal models [10, 11]. Recent developments in imaging technology have made it possible to use the diffusion tensor magnetic resonance imaging (DT-MRI) method to examine the distribution of fibers in the human body [12]. However, it is difficult to recreate such a fiber tissue paradigm in a computer framework for four reasons. Before they could be merged with a tissue matrix volume, the fibers in a DT-MRI model were first represented as lines or splines, which needed to be translated to volumes. Second, it is quite challenging to create precise fiber meshes that may be employed in studies since a significant portion of fibers overlap one another. Third, it is challenging to determine the precise number of fibers in a place without first conducting a histological analysis to determine the fiber volume percentage (FVF). Fourth, the majority of tissues continuously blend with one another without any obvious boundaries. For instance, it is challenging to isolate the left ventricular (LV) tissues of the heart as they change into the right ventricle (RV) and other heart chambers. There have not been many attempts to include tissue anisotropy in finite element (FE) models in the literature. The most typical approach has been to roughly identify a primary fiber path in each discretized region of a tissue or organ (with discernable fiber orientation). In comparison with other orientations, the primary fiber direction is ascribed a stiffer material attribute [13], and this also enables loading via a variety of passive excitation techniques [14]. Modeling a tissue region using a transversely-isotropic material formulation has been another approach [15, 16].

8.3.4 Handling and Management of Tissues

To progress translational research aimed at identifying and describing methods for individualized (personalized) medical care, the accessibility of human tissues to assist biomedical research is essential. In order for modern medical research in fields like cancer and diffuse lung disease to advance, researchers ultimately need access to high-quality specimens of human tissue, including physiological fluids. In order to assist researchers in biomedical research, numerous organizations are now engaged in the collection, processing, storage, and dissemination of human tissues. To help biomedical research, there are several ways to collect human tissues and physiological fluids. The least organized of these is best characterized as "catch as catch can," in which surgeons, pathologists, or other medical professionals deliver tissue specimens to biological researchers as they become available. These samples are typically only collected when there is time and when the order is recalled. Standard operating procedures, for example, were not used to collect, process, or store these specimens; thus, their quality may be poor, and the assessment may be incorrect. Typically, these samples do not have any quality control.

Purchasing research-grade human tissues from a tissue organization that uses a "banking model" is a more organized method. A standardized operating procedure (SOP) is adhered to in the banking model to gather, process, and preserve human tissues. Such tissues are frequently frozen or gathered and turned into paraffin blocks. A tissue bank normally does not offer fresh and unfrozen samples or "specially processed" tissues. Therefore, the primary drawback of the banking approach is that the samples might not satisfy the needs of the investigator in terms of size, processing, and storage, and needed normal tissues or fresh tissue may not be available. The "banking model" has the benefit of allowing researchers instant access to huge numbers of specimens as well as associated clinical and demographic data, including clinical outcome.

8.3.5 Ethical Issues with in Vivo Testing

Sometimes, the human tissue samples are quickly processed and frozen after the surgery. While a few tissues can be acquired, processed, and frozen within 15 min following surgical removal. Although most tissue repositories lack the personnel and resources necessary to rapidly collect and process a specific type of tissue, it is nonetheless crucial to reduce the time that passes between the extraction of operative specimens and their transport from the operating theater to the tissue repository. The time before preservation should be as low as possible, but processing delays may happen when a particularly large specimen or several specimens need to be processed at once. In this case, one or two aliquots from each specimen can be quickly frozen in liquid nitrogen or on dry ice, and further aliquots can be gathered and frozen at a later time.

Also, researchers might not have easy access to ultra-cold storage devices for preserving the tissue specimens. Tissue repositories may momentarily hold tissue samples for researchers, but they should charge for this storage. The best way to store tissues for at least six months in order to facilitate scientific research is still up for debate. Unprocessed tissue specimens should not be kept at temperatures below 20 °C for longer than one month, and the tissues should never be kept in cryostats or self-defrosting freezers for an extended amount of time.

References

1. Budday S, Sommer G, Birkl C, Langkammer C, Haybaeck J, Kohnert J et al (2017) Mechanical characterization of human brain tissue. Acta Biomater 48:319–340. https://doi.org/10.1016/j.actbio.2016.10.036
2. Rashid B, Destrade M, Gilchrist MD (2012) Mechanical characterization of brain tissue in compression at dynamic strain rates. J Mech Behav Biomed Mater 10:23–38. https://doi.org/10.1016/j.jmbbm.2012.01.022
3. Ghajar J (2000) Traumatic brain injury. Lancet 356:923–929. https://doi.org/10.1016/S0140-6736(00)02689-1
4. Navindaran K, Kang JS, Moon K (2023) Techniques for characterizing mechanical properties of soft tissues. J Mech Behav Biomed Mater 138:105575. https://doi.org/10.1016/J.JMBBM.2022.105575
5. Hu T, Desai JP (2004) Characterization of soft-tissue material properties: large deformation analysis. Lect Notes Comput Sci (Including Subser Lect Notes Artif Intell Lect Notes Bioinformatics) 3078:28–37. https://doi.org/10.1007/978-3-540-25968-8_4/COVER
6. Polio SR, Kundu AN, Dougan CE, Birch NP, Ezra Aurian-Blajeni D, Schiffman JD et al (2018) Cross-platform mechanical characterization of lung tissue. PLoS ONE 13:e0204765. https://doi.org/10.1371/JOURNAL.PONE.0204765
7. Picinbono G, Delingette H, Ayache N (2001) Nonlinear and anisotropic elastic soft tissue models for medical simulation. In: Proceedings of the 2001 ICRA. IEEE international conference of robotics automation (Cat. No.01CH37164). IEEE. https://doi.org/10.1109/ROBOT.2001.932801
8. Holzapfel GA (2000) Biomech preprint series biomechanics of soft tissue
9. Chanda A, Unnikrishnan V, Roy S, Richter HE (2015) Computational modeling of the female pelvic support structures and organs to understand the mechanism of pelvic organ prolapse: a review. Appl Mech Rev 67. https://doi.org/10.1115/1.4030967/370016
10. Lowry OH, Gilligan DR, Katersky EM (1941) The determination of collagen and elastin in tissues, with results obtained in various normal tissues from different species. J Biol Chem 139:795–804. https://doi.org/10.1016/s0021-9258(18)72951-7
11. Neuman RE, Logan MA (1950) The determination of collagen and elastin in tissues. J Biol Chem 186:549–556. https://doi.org/10.1016/s0021-9258(18)56248-7
12. Basser PJ, Pajevic S, Pierpaoli C, Duda J, Aldroubi A (2000) In vivo fiber tractography using DT-MRI data. Magn Reson Med 44:625–632. https://doi.org/10.1002/1522-2594(200010)44:4
13. Li X, Kruger JA, Nash MP, Nielsen PMF (2011) Anisotropic effects of the levator ani muscle during childbirth. Biomech Model Mechanobiol 10:485–494. https://doi.org/10.1007/s10237-010-0249-z
14. Colli Franzone P, Guerri L, Pennacchio M, Taccardi B (1998) Spread of excitation in 3-D models of the anisotropic cardiac tissue. II. Effects of fiber architecture and ventricular geometry. Math Biosci 147:131–71. https://doi.org/10.1016/S0025-5564(97)00093-X
15. Almeida ES, Spilker RL (1998) Finite element formulations for hyperelastic transversely isotropic biphasic soft tissues. Comput Meth Appl Mech Eng 151:513–538. https://doi.org/10.1016/S0045-7825(97)82246-3

16. Weiss JA, Maker BN, Govindjee S (1996) Finite element implementation of incompressible, transversely isotropic hyperelasticity. Comput Methods Appl Mech Eng 135:107–128. https://doi.org/10.1016/0045-7825(96)01035-3